pu771n' 4 S70Pp3R 1n 73h Num83rS 0F D347h

&

The Occult

by

gary of the house of Fraughen

ISBN – 978-1-8382745-1-1

This paperback edition, published 2022

Contents

Preface	Pg 1
Bill Hicks – LSD	Pg 4
Chapter 1 – Numbers, Pyramids and Patterns	Pg 5
Chapter 2 – Pizza, Time, Space and 9	Pg 17
Chapter 3 – Numerology and Gematria	Pg 23
Chapter 4 – Numbers	Pg 27
Chapter 5 – Coronavirus, Shakespeare and Faces	Pg 67
Chapter 6 – Roulette, Snakes and Squid Game	Pg 89
Chapter 7 – The Sun, Sun Dials and Rolex	Pg 97
Chapter 8 – Shipping, Serial Killers and Adrenochrome	Pg 107
Chapter 9 – Magic Squares, Turtles, and Demons	Pg 115
Chapter 10 – Computers, Parachutes and Nuclear War	Pg 133
Chapter 11 – Music, Pyramids and Tesla's 3, 6 and 9	Pg 139
Chapter 12 – Body Parts, Distance and the Soul	Pg 151
Chapter 13 – Masonic Ciphers and a Code to Crack	Pg 157
Chapter 14 – Hate Numbers, Hanging and Jeans	Pg 163
Chapter 15 – Shoes Laces, DNA and Death	Pg 167
Chapter 16 – Movies, Easter Eggs and Milk	Pg 169
Chapter 17 – Giants and Champagne	Pg 189
Chapter 18 – Blood, Zombies and Plimsoles	Pg 193
Conclusion	Pg 207
Gematria Keys	Pg 209
Further Research	Pg 211

Preface

Beads before the thread.

We are all entities on a blue planet, a crystal glassy bright gemstone set into the cavernous jet-black expanse of an aggressively hostile deep cold solar system.

Terra Firma, some 4.6 billion years old, sustaining at present over 9 million species of life with nearly 7,800,000,000 of them genetically linked to one species. A collective hurtling and rotating on its planet's axis around the sun, through space at 67,000 mph while immersed in culture of the game.

A game of scrambling around, trying to find an edge. The challenge of making the most of what each of us are presented with each day. An age of going through a system of moneylending, spending, and usuries.

What if I were to tell you that you were dying would your view on gaining an edge that day change? Would your view of money change? Would your goals change? Would you change, would your need for scrambling around each day change? Would the game change?

Question...why not just change the game right now, while you are healthy and alive, while you still have plenty of time. Why you need near death to make you change.

A wise man once said:
"Once the mind is stretched by a new idea, it never returns to its original dimensions".

Greetings
You have come back; I am incredibly pleased. I shattered many of you like falling bone china on terra-cotta in the *stopper 1*, with words. Now we are going to do it again... with numbers. Before I raised a fire in your mind's eye, now we fan the flames.

Those who have not read my first attempt in wrestling with writing a book no matter, this second one will stand in good stead on its own. There are, however, elements in the *stopper 1* that will add to our fun when we dive deeper into numbers, but it is not essential. As before, like a gladiator, I will continue to entertain, but you will have to put in some work. You will have to put this book down and do things, find things, confirm things, calculate things and there may even be some financial benefit. But most of all we will have fun.

The titles of my books so far are called *"putting a stopper in the bottle or numbers of death"* very odd I admit, but my reasoning is to re-fill the gaps in your existence as a bottle can be emptied and refilled. To unlearn and relearn. To reveal things right in front of you of which you cannot see which is hidden in plain sight. To stopper that bottle once more and not allow others to pollute its contents with deception, when replenished. It then becomes yours once more but cleansed with knowledge. There are occult (things hidden) used on us in our everyday lives and no more so than with numbers.

Many of you had kind words for me, some of you I have met, while others have emailed me or penned me letters to my home. People have contacted me to say they had not read a book since childhood, and they found *stopper 1* fascinating with the oldest of this collective being 83. It has been a true demonstration of how humans can appreciate each other and to those, I thank you.

When I started my working life as a labourer, I could never have imagined that by 55 I would be finally enjoying my existence. I can declare I have always worked hard but it took a long while to work smart, both directions were wrong, it is about doing what you enjoy. I have figured (number term) out that rather than die with a large house full of furniture, I am going for a passport full of stamps surrounded by kindness. Since the publication of *stopper 1* we have suffered more of the chant of 3 through Fear, Fear, Fear. The Mass... Formation... Psychosis... with:

The Great Reset
The Great Awakening
New World Order
One World Government
Test and Trace
Delivering Peoples Priorities
Together We Win
Build Back Fairer
Build Back Safer
Build Back Better
Keep Life Moving
Trust the Science
Get the Jab
Copy, Acquire, kill
Hands, Space, Face
Nobody Left Behind
Global Information Control
Trusted News Initiative
Wash hands, Cover Face, Make Space
Stay Home, Protect the NHS, Save Lives
The New Normal
Get Boosted Now... the number 3 in neurolinguistic programming. These spells do not work on you anymore if you read *stopper 1* and our tribe of the awakened grows every day.

In this work we' explore numbers across many levels due to people wanting more in this specific area. Human history in my view is false, His-story is not our Story. I believe humanity, historically was far more in tune with nature, the geometric universe, time, distance, measurements, language, and numbers. I invite you to explore these with me and will point out specific occultic numbers that surround us every day with things that have been kept from you. The master builders, mystery schools and wisdom holders of old used these numbers throughout life and these have been removed from our understanding.

To encrypt something is a simple means of hiding the code and only those able to decipher the information will understand the information.

We live in a mathematical universe which is a fingerprint of creation and numbers are on everything, and I will prove it.

If you have come back for more after reading *stopper 1* you are a free-thinking individual, who, is able to critically look at a narrative and ignore any influencing factors using your own perspective. To question the narrative and arrive at your own conclusions, it means you are like me and vice versa, we are different from the majority.

As before I am going to continue to spell things out for you, things that most people never see. *Numbers* will drive these pages of occult. Are you sitting comfortably, if so relax, I require your mind to "open" like a parachute otherwise this will fail to work.

So once again, let us start our little journey...just you and me...in what time you can spare...here and there.

Gary, Just a builder.

That which you cannot see, can cause you harm.

Bill Hicks

“Wouldn't you like to see a positive LSD story on the news? To base your decision on information rather than scare tactics and superstition? Perhaps? Wouldn't that be interesting? Just for once”?

"Today, a young man on acid realized that all matter is merely energy condensed to a slow vibration, that we are all one consciousness experiencing itself subjectively, there is no such thing as death, life is only a dream, and we are the imagination of ourselves. Here's Tom with the weather".

1961 – 1994, RIP. Died aged 32.

Chapter 1 – Numbers, Pyramids and Patterns

Ready...steady...go...but wait...

do I mean 3...2...1...go? Or should that be 1...2...3...go.

Numbers can camouflage words and words can be numbers, so we start with confusion as an opening gambit.

The word number is from the 1300s (from the Latin *numerus* sum aggregate of a collection) but as a word it is complex for it can be re figured into numbers, numbered, numberer, numbering, prenumber, renumber, numberless, mis-numbered, and outnumbered.

Have you ever considered the words we use involving numbers, if not, here are a few:

A bunch of fives, a stitch in time saves nine, at sixes and sevens, back to square one, better half, catch 22, cloud 9, dressed to the nines, fifteen minutes of fame, counting sheep, fifth column, goody two shoes, half hearted, million-dollar question, four eyes, forty winks, give me five, a trick or two, nine 2 five, hung drawn and quartered, no quarter given, penny for your thoughts, one stop shop, one hit wonder, seven-year itch, the whole nine yards, third degree, third time lucky, two cents worth, penny's worth, zero tolerance and even in foreign terms, double Dutch or menage a trois.

Three sheets to the wind, three strikes, three ring circus, three Rs, three squares a day, even three cheers and you are 6 feet under when your, number is up. Specific numbers invade sentences like six pack, six of one and half a dozen of another, six ways from Sunday.

While most people understand what a number is, they do not actually study them. You can walk in the garden or forest without knowing a thing about photosynthesis. You can watch the sunrise and sunset without knowing about the rotation of the Earth or plain or gaze upon the stars without knowing what a parsec is. Occultists can have many disciplines, some words some numbers and some sigils from different secret societies. I believe in the free pursuit of knowledge that is for everybody, mystery cults and schools do not agree with me.

I have determined the reasons they keep this hidden knowledge from the masses is for the following reasons:

1. Proof of dedication and worthiness.
2. Keeping sensitive information from those who would abuse it.
3. Keeping the specifics of rituals and spiritual experiences secret so that initiates are blind to it and experience it themselves in stages.

So let us explore this hidden world of numbers that exists.

The word 'figure' can also be a number but look again and you see this figure or that figures to figure something out, it figures. So now you see the confusion between words and numbers and how they both can be the same. Figure can also be a human shape.

We live in an existence like animals that are kept on a farm, penned in, isolated from those who are different. Life is about experiences and the journey is as important as is the destination. The controllers of the farm do not want us to see our real existence. As a collective we are one, but the purveyors of power have stepped in between you and the creator (whatever the word creator means to you). We have been conned into rules, laws, acts, statutes, licences, and certificates all with their own numbers, which have been designed to lock us into a lower vibrational state (frequency numbers). My take on our present existence is this:

"History is a lie, religion is a control system, money is a hoax, debt is a fiction, media is manipulation, government is a corporation, and numbers are hidden. You are living in an illusion based on words or it may even be a matrix of numbers that you have never seen".

We communicate using words, but do not realise it is only half of the skillset available to us. Mathematics is the other half, yet we ignore it. As you read this work, I will on occasion write the numbers numerically not alphabetically; ONE will become 1 to allow your mind to see their existence as you read.

Writing actual numbers in English grammar should never be used in a written text, but this was never the case historically. Numbers themselves were stand-alone symbols and were part of written sentences but this has been changed in the last few decades. Numbers and words have become 2 separate languages, as words in paragraph form, and numbers into spreadsheets like news was in broad sheets (newspapers). Consider this fact, why do we need the written word such as Thirteen when we have 13? There is quite simply no need for it.

Spoken numbers seem logical, but are they? From 21 to 99 the numbers are logical, but the words are not. Eleven and Twelve come from the old English words *endleofan* and *Twelf,* traced back further to a time when they were *ain+lif* and *twa+lif.* So, what does this *lif* mean? Etymologically terms it means "to leave". *Ainlif* is *"one left (after ten)"* and *twalif* is *"two left (after ten)".* So, the question is, why don't we have *threelif, fourlif, fiflif, sixlif* and so on? When I write technical documents, I use a book called the "Chicago Manual of Styles" and the *rule of thumb* (1 inch which I will explain later) declares that all numbers from 1 to 9 should be in words and any number from 10 onwards can be numeric although not essential. I point this out because it is fundamental, yet most people have never even thought about it. We created words for 11 and 12 a long time ago by calling them "one left after ten" and "two left after ten". They were more useful to us than the higher numbers. Irregularities of pronunciation appear in the tens (*twenty, thirty, fifty* instead of *twoty, threety, fivety*) because we have been making everyday use of those numbers for longer than we have for *two hundred, three hundred,* and *five hundred*). *Thousand* is an old word, but its original sense was "a great multitude".

My analysis is that it is to move us away from the Imperial System and steer us into the Metric System. For example, 1 ½ inches appears cumbersome as opposed to writing one and a half inches which, was not the case 70 years ago, we have merely been reprogrammed into this new Metric System. The reasons for this occult I will get to later.

Within the book I will visit some aspects of religion, so for those who do not have a propensity in this form of belief I apologise. I am raising these more from a spiritual nature as I believe these holy books are a warning to humankind. Each religion has God in a descriptive nature, but I believe each book is just a unique way of recognising a belief system of creation. We have oceans, lakes, rivers, pools, ponds, and streams but they are simply different words describing the same thing which is WATER. All organised religions have several ways of describing the source code of creation, and while being allegory it is also hiding numbers.

Geometric and mathematical patterns govern life, time, and existence. Sounds a bit heavy as an opening statement of fact but without numbers you simply would not exist as an entity in your present form. I intentionally include the word *sound* by no mistake as this is energy vibrating within mathematics.

As a Chartered Building Surveyor, I drown in a sea of numbers including Stadia, Levels, Datums, Elevation, Maize field, Trig, Azimuth, Psychometrics, the list is exhaustive. You know nothing of my profession and I of yours because each of us have not been taught each other skill sets. There is, however, a common denominator and that is the mathematics that was never taught in schools. It is occulted or hidden from you. How can that be I hear you cry but simply we all swim in a soup of numbers, but they have a power that has never been pointed out to you. We often hear the phrases "he has a sense of humour", or "sense of rhythm" but mathematics is a sense of patterns and I intend to reveal them. Numbers to me are patterns, some have rules, others seem random, some numbers have rhythm, some regularity, structure, or repetition. I am not a numbers nerd but embrace numbers the same as I do letters because both have hidden codes.

In Western society we are ashamed of saying we cannot read but we take a sense of glee by saying *I have no head for numbers*. This declaration really means I only speak half our communicable language. We learn foreign languages but in mathematics we do not learn foreign ways of calculation, which sounds insane but this I will show you in a later chapter.

Numbers dictate how we run our lives, they tell us when to wake up and when to sleep, they tell us where to go and how long it will take to get there. Numbers tell the truth, and they also lie. Numbers are lucky charms but also things to avoid. They are the key to understanding the universe, distance, motion, time, measurement, birth and even death. They can even repeat themselves to us and people often see the repetition of a number set. Used by nature they are simply embedded in the fabric of our existence. They can be a tool of fact and conversely manipulation and deceit.

Superstition used to run our lives, then it was religion but now it is numbers and science. This year all we have heard is *"trust the science"* and for those who follow my work it is a chant of 3. They are concepts of simplicity and complexity visible and hidden entwined in the pinched-up matter of carbon with our human bodies.

The fear of numbers is called numeraphobia, but I believe this stems from the fear of going to school which is called didaskaleinophobia. When I talk to people who dislike numbers all seem to have had some sort of unpleasant experience as a child or in a classroom setting with numbers which seems to have left long-lasting negative and deep-seated anxiety. This problem runs deep as there are words for phobias in relation to every single number in existence from the fear of number 1 henophobia (Greek – *fear of one*/ Latin -*Unus*) to infinity and I will point some out. The fear of large numbers is meganumerophobia or odd numbers imparnumerophobia and this is consistent with PI, golden ratio, prime numbers, irrational numbers, negative numbers, and even decimals. Ask yourself why there is a different phobia for every single number in existence. I believe the state school system and mainstream curriculum intentionally frightens children to keep them away from numbers and therefore half of our communication. After all, if you do not know how to deal with tax, debt, interest, or money, you end up a slave to a number language which you know nothing. My intention in these pages is to show you the world of numbers that exist beyond your wildest imagination.

People who investigate numbers in my mind are 'explorers' of another realm and these include mathematicians, astronomers, physicists, hackers, gamblers, economists, statisticians, and engineers. I use the word explorer because they discover things...things that have always been there, but they have just not been obvious. I believe that numbers are ghosts in the matrix until discovered.

Numbers are created twice, once by the universe and then by us as humans when we discover them and manifest them into our known existence and language.

We live on a consumption planet which is why we are called 'consumers' from the smallest microbes to the largest whale, something is eating something else to go on living. Our very existence relies on the fact that more numbers or less numbers which means more food or less food. Not every organism counts the numbers, but they undoubtingly recognise the pattern of them. More food we survive, less food we die.

Imagine we collect pebbles, the more pebbles you have the larger the number which seems obvious but when you get into double and triple digits keeping track of the amount becomes an issue. A small bag of pebbles is different to a medium or a large bag and then the number of pebbles becomes important.

Where do numbers come from?

Humans are bipedal (terrestrial locomotion using 2; with 2 legs, 2 arms, 2 feet, 2 eyes and 2 parts to our brain). Each of our 2 hands consists of 4 fingers and 1 thumb or in number terms 5. The *Sumerian Texts,* the oldest stone carvings still available to view in our own known human history has calculations based on 5, so you can recognise the correlation.

Hammurabi the oldest numbered codes were from Babylonia carved onto stone (*stele*) discovered in Susa Iran in 1901 and said to be dated 1792-1750 BC. The 282 laws included slavery, women and children's rights, property, farming, and monetary rules. The sentences presumed innocence in proving the breaking of such laws and penalties could never exceed the rule breakers actions. Stone carvings were difficult but notches in bones using flint became the easiest form of recordkeeping for sea traders.

In *stopper 1*, I visited the occult of language but with numbers there is a crossover. In the West we take it for granted that we write from left to right, but with numbers we work from left to right, top to bottom (fractions), bottom to top (equations) and sometimes from right to left (string theory). Many people contacted me wanting to know more so I spent some time laying out a family tree of letter direction and where our languages emanated from.

I have broken these directional origins down into 5 different areas and they are explained as follows:

- Alphabet
- Abjads
- Abugidas
- Featural Alphabet
- Relief

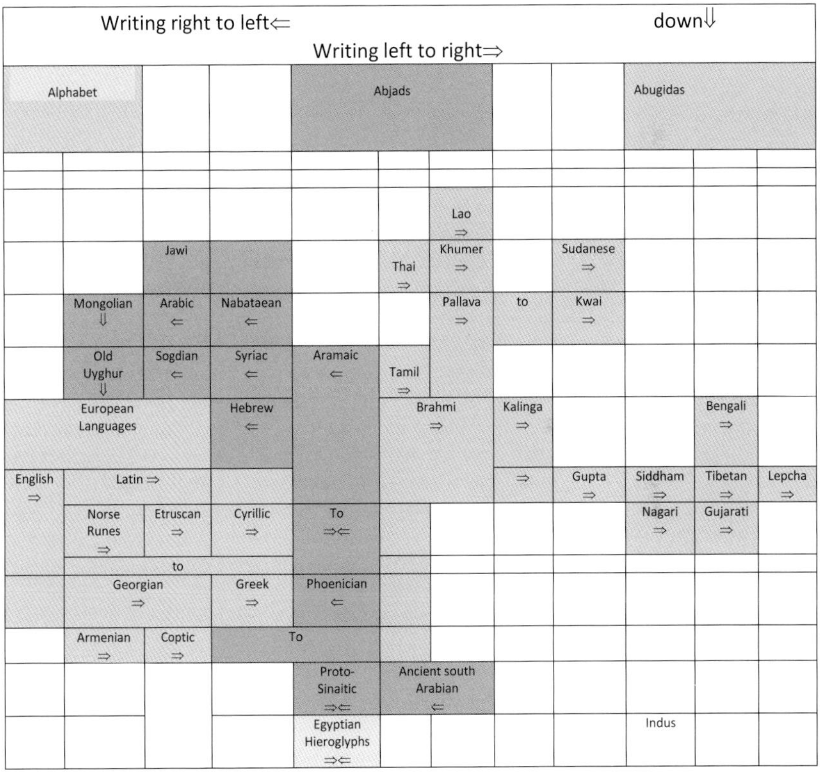

Writing right to left⇐
down⇓
Writing left to right⇒
Alphabet
Abjads
Abugidas
Lao ⇒
Jawi
Thai ⇒
Khumer ⇒
Sudanese ⇒
Mongolian ⇓
Arabic ⇐
Nabataean ⇐
Pallava ⇒
to
Kwai ⇒
Old Uyghur ⇓
Sogdian ⇐
Syriac ⇐
Aramaic ⇐
Tamil ⇒
European Languages
Hebrew ⇐
Brahmi ⇒
Kalinga ⇒
Bengali ⇒
English ⇒
Latin ⇒
⇒
Gupta ⇒
Siddham ⇒
Tibetan ⇒
Lepcha ⇒
Norse Runes ⇒
Etruscan ⇒
Cyrillic ⇒
To ⇒⇐
Nagari ⇒
Gujarati ⇒
to
Georgian ⇒
Greek ⇒
Phoenician ⇐
Armenian ⇒
Coptic ⇒
To
Proto-Sinaitic ⇒⇐
Ancient south Arabian ⇐
Egyptian Hieroglyphs ⇒⇐
Indus

Humans still to this day write letters and numbers in 3 directions depending on geography and culture, the numbering systems follows the direction of their letters but there was however a 4th.

Boustrophedon carvings on ancient stone worked both right to left and left to right with the first line being left to right then it snakes back around, and the next line below is right to left and so on. The word *boustrophedon* means "as the ox ploughs" as a method of writing. Each line could be spaced perfectly beneath the one above, therefore making the relief work easier to control for the stone mason. As numbers were letters it was easy to place these within the text or relief. This system, however, is no more.

Culturally numbers followed the direction of work, all the calculations follow the direction of text. Only as mathematics gets more complicated does it start changing direction.

The Metric System has also changed all this as calculations work both vertical and horizontal.

The Romans perfected notching and created the symbology of cuts as letters and numeric presentation was to come later.

The letters they used	I	V	X	L	C	D	M
Represented	1	5	10	50	100	500	1000

This created a problem because there was no zero or 0 for it was in letter form and not numeric. Remarkably the Babylonians, Mayans, Greeks, and Chinese struggled because they did not have zero/0. This was followed by the abacus or counting frame which was a form of positional notation. At about the same time in India a well-respected astronomer called Brahmagupta created the concept of zero or 0 and the game changed in 628 A.D. Works from Brahmagupta (630 AD) ended up in the writings of Al Khwarizmi (820 AD) whose name later became known as Algebra.

In the West, the work by Leonardo of Pisa who became known as Fibonacci was another game changer. Within his Latin writings he recorded "*Nouem Indorum he sunt 987654321 cum his itaque nouem figuris, et cum hoc signo 0, quod arabiee zephirum appellatum scribitur quilibet numerus, ut inferior demonstrator*". It introduced modern numeric and the zero as a character representation translated it reads "here are the nine Indian figures". Interestingly if you note he calculates down from 9 to 1 when we read it left to right but because in Arabic you read from right to left you can now see the 1 to 9 in descending order. It is also worth noting that *zephrium* or *Zephyr* is within this statement which is where the term *cipher* originates.

This led to the decimalisation and numbers 0, 1, 2, 3, 4, 5, 6, 7, 8, 9, 10.

Angles, Numbers and Dots

Have you ever thought about the shape of numbers themselves? Let us unlock this as a first step; now we move to numeric representation of numbers recognised today which have pyramids running through them in occult. We will explore this codex system of pyramids on paper.

1 is simple as it is singular, it stands alone and originally was I, do you notice the original I is different to 1? A zero or 0 has no angles which meant no number 0 by placing a dot into the triangle pyramid system it leaked into the numeric presentation. Zero meant no pyramid and the subsequent number before 0 increased the total.

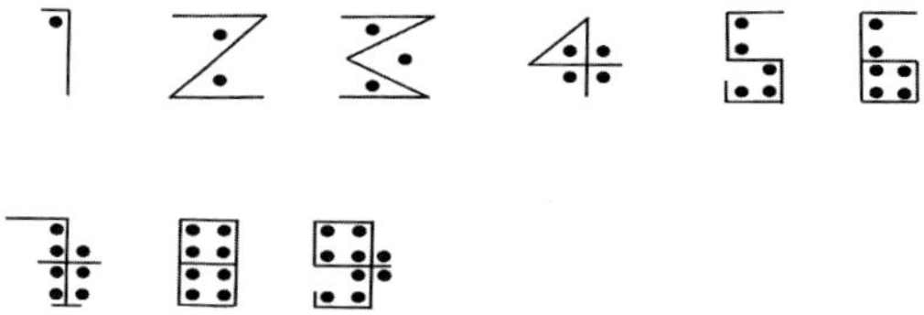

Now you are aware that numbers are really hidden pyramids, but these pyramids run deeper with number patterns. Humans have 3 eyes 2 to look and 1 to see or realise. Patterns in the numbers are essential to cracking codes.

The pyramid lays at the heart of the mystery schools but where did this triangle come from.

Using the number 10 from the Commandments of God (Georgia Guidestones) this can become a pattern that can be inverted, let us explore the pyramid itself first.

If you lay an equal distance pyramid with 1s you get the following pattern:

```
   1
  1 1
```

Next 6x1s

```
   1
  1 1
 1 1 1
```

Then 10x1s

```
   1
  1 1
 1 1 1
1 1 1 1
```

This then continues with 15 of the number 1 and 21, 28, 36, 45, 55 etc. But now you start to see the numbers in the pattern.

```
1              =1
1  1           =2
1  1  1        =3
1  1  1  1     =4
-------------
4  3  2  1
```

Here is some more number combination into the patterns.

1
1 1
1 2 1
1 3 3 1
1 4 6 4 1
1 5 10 10 5 1
1 6 15 20 15 6 1
1 7 21 35 35 21 7 1
1 8 28 56 70 56 28 8 1
1 9 36 84 126 126 84 36 9 1
1 10 45 120 210 252 210 120 45 10 1
1 11 55 165 330 462 462 330 165 55 11 1
1 12 66 220 495 792 924 792 495 220 66 12 1
1 13 78 286 715 1287 1716 1716 1287 715 286 78 12 1
1 14 91 364 1001 2002 3003 3432 3003 2002 1001 364 91 14 1

The above is known as Pascal's triangle. Simply put, each number is followed beneath with the simple rule that the number must add up to the two numbers above. Patterns born from specific numbers:

Number 1

1x1=1
11x11=121
111x111=12321
1111x1111=1234321
11111x11111=123454321
111111x111111=12345654321
1111111x1111111=1234567654321
11111111x11111111=123456787654321
111111111x111111111=12345678987654321

Number 4

16×4=604
166×4=664
1666×4=66664
16666×4=66664
166666×4=666664
1666666×4=6666664
16666666×4=6666664
16666666×4=66666664

Number 8

1×8+1= 9
12×8+2=98
123×8+3=987
1234×8+4=9876
12345×8+5=98765
123456×8+6=987654
1234567×8+7=9876543
12345678×8+8=98765432
123456789×8+9=987654321

Number 9

0×9+1=1
1×9+2=11
12×9+3=111
123×9+4=1111
1234×9+5=11111
12345×9+6=111111
123456×9+7=1111111
1234567×9+8=11111111
12345678×9+9=111111111

9×9+7=88
98×9+6=888
987×9+5=8888
9876×9+4=88888
98765×9+3=888888
987654×9+2=8888888
987 6543×9+1=88888888
9876 543 2×9+0=888888888

9x1=9
9x2=18;1+8=9
9x3=27;2+7=9
9x4=36;3+6=9
9x5=45;4+5=9
9x6=54;5+4=9
9x7=63;6+3=9
9x8=72;7+2=9
9x9=81;8+1=9
9x10=90;9+0=9
9x11=99;9+9=18;1+8=9
9x12=108;1+0+8=9
9x13= 117;7+1+1= 9
9x14=126;6+2+1=9
9x15=135;5+3+1=9
9x16=144;4+4+1=9
9x17=153;3+5+1=9
9x18=162;2+6+1=9
9x19=171;1+7+1=9
9x20=180;8+1+0=9

12345679×9=111111111
12345679×18=222222222
12345679×27=333333333
12345679×36=444444444
12345679×45=555555555
12345679×54=666666666
12345679×63=777777777
12345679×72=888888888
12345679×81=999999999

Number 37

37×3=111
37×6=222
37×9=333
37×12=444
37×15=555
37×18=666
37×21=777
37×24=888
37×27=999

A mirror pattern using all 9 numbers

987654321	123456789
087654321	123456780
007654321	123456700
000654321	123456000
000054321	123450000
000004321	123400000
000000321	123000000
000000021	120000000
000000001	100000000
-----------------+	------------------
1083676269	1083676269

Below is a *magic square* which I will explain in another chapter, but every horizontal line, vertical line, and diagonal line add up to the same number 111. Additionally, this *magic square* has other hidden numbers for example 6 rows and 6 columns, and you realise 6x111=666.

6	32	3	34	35	1	= 111
7	11	27	28	8	30	= 111
19	14	16	15	23	24	= 111
18	20	22	21	17	13	= 111
25	29	10	9	26	12	= 111
36	5	33	4	2	31	= 111

In history there have been quoted sentences about numbers that have resonated such as:

"*Nature is written in mathematical language*" - Galileo

"*If you only knew the significance of the 3, 6 and 9 then you would have a key to the universe*" - Nikola Tesla

"*It is forbidden to kill; therefore, all murderers are punished unless they kill in large numbers and when they do this there is the sound of trumpets*" - Voltaire

"*You have to be odd to be number one*" Dr. Seuss

Magic
As children we all know about the word *abracadabra* which is a child's way of casting a spell in words, there is however, a numeric value based on the pyramidal system. If we look at the word abracadabra its origins stem from ancient aromatic in the use of sorcery, wizardry, and

witchcraft. It is well laid out in numeric form; this is how abracadabra used to appear on amulets and tablets of that time. For now, remember the pattern, as we are covering the basics as we unpick the occult.

A - B - R - A - C - A - D - A - B - R – A	A
A - B - R - A - C - A - D - A - B - R	A - B
A - B - R - A - C - A - D - A - B	A - B - R
A - B - R - A - C - A - D - A	A - B - R - A
A - B - R - A - C - A - D	A - B - R - A - C
A - B - R - A - C - A	A - B - R - A - C - A
A - B - R - A - C	A - B - R - A - C - A - D
A - B - R - A	A - B - R - A - C - A - D - A
A - B - R	A - B - R - A - C - A - D - A - B
A - B	A - B - R - A - C - A - D - A - B - R
A	A - B - R - A - C - A - D - A - B - R - A

In Greek, the Kabbalistic word for ABRAXAS symbolises the soul year of 365 days and 365 heavens because the sum of the value of the Greek letters add up to 365. The spelling originates in the confusion made between the Greek letters Sigma and Xi and in the Latin translation. In antiquity the word Yahweh or Jehovah could never be mentioned as they were words for the supreme being. Abraxas took its place.

In etymological terms *Abracadabra* means "*I will create when I speak*". There is, however, an inversion but if the amulet was turned with a pyramid point vertically then the spellcasting would use the term *Avada Kedavra* which translates to "*I will destroy as I speak*". There is, however, a positive in both symbols. If you wrote this pyramidal downwards it would rid yourself of evil but the inversion of the same pyramid or upwards was regarded mischief, in the ancient world people wore the left version as a constant for either keeping out evil or for good luck. Conversely bad luck spell casting is the right.

Each side of the sigil totals is 11
So 11,11,11
Or 2+2+2=6
Or 11+11+11=33
Or 1+1+1+1+1+1=6
Or 111+111=222

Pythagoras and his mystery school considered the triangular pyramid as a *Monad* (1-male) also involves *Dyad* (female). The left version was male, the right version was female as the depiction represented a hidden form of genitalia. Interestingly dyad means 2 and this belief system was that a monad represented ORDER while the dyad represented DISORDER. The word disorder in this context, however, was not considered a negative for a male (singular) and can only increase his existence with an addition thereby creating 2 and 3 and a new-born baby can create disorder but not in a negative context.

These mathematical numbers of concepts are not wired into the human condition, but patterns are. On 13th March 2017, Associate Professor Caleb Everett from the University of Miami with other Anthropologists worked with the indigenous Amazonian people known as the Pirahã. During their investigation, it was realised that these people had no identifiable language for numbers. The discovery is that if they were presented with 3 things on a table (in this case batteries) they could replicate the amount presented. There was, however, a problem which was number 3 and above, anymore and they were going to make mistakes. The researchers realised that extraordinarily because they had no words for numbers and mathematical concepts were not wired into the human condition, linguistic transmission prevented them any ability to count. If, however, these batteries were presented in a pattern then they could replicate the number of batteries on the table.

Which of these has four zero and three four?

A 0000444

B 4034

C 000034

All of them, but the language in relation to the numbers is diametric. Numbers are multidimensional in terms of the form, function, and presentation. Now you understand the reason I may write the numbers in this book numerically not figuratively.

In the past numbers were written as letters. People needed to count but had no need to write it in words. The word ILLITERACY relates to the inability to process letters which are constructed into words. Illiteracy with numbers is called INNUMERACY but this has no relation to pattern recognition. My analysis is that the mystery schools and ancient holders of knowledge recognised this most significant difference. You will need to consider this as you work through Sigils and Ciphers in other chapters.

Throughout this book remind yourself constantly that numbers sit behind the material universe in 4 ways.

1. Pure Numbers in Arithmetic
2. Time and Harmonics or Audio
3. Space and Geometry
4. Time and Space Celestial Astronomy and Physical Anatomy

At the end of this chapter, you may feel a little confused, but I am taking you on a journey and first I must lay out the reasoning behind why things are done in the mystery schools. The recognition of a pattern is like watching an ECG of a heartbeat and if you see the rhythm (pattern), you can understand if something is healthy or unhealthy. So, remember *ABRACADABRA* and its patterned layout, it will suddenly make sense in a later chapter.

Chapter 2 - Pizza, Time, Space and 9

In our daily lives' numbers reveal much more than people realise and do not see, so let us start with a basic introduction regarding space and time.

Pizza

Shapes can appear complicated, but I declare this as a statement, "*that the universe can be described as pizza*". You must think I am crazy but indulge me. This carbohydrate snack holds the secrets to the universe. It is made flat and round, sold in a square box, and eaten as a triangle, but it is the numbers in these 4 shapes that need your attention.

Space Measurement

Physical measured space has its own set of numbers. 1 foot is 12 inches, 1 sq. foot is 144 sq. inches and here we start the correlation as 24 hours x 60 minutes is 1,440 minutes and the 144 appears as with time. Coincidence? There is no such thing.

Then we have a cubic foot which is 1,728 cubic inches and 1 square yard is 1,296 inches, interestingly if we add a zero, we get 12,960 but why is that important?

Well in geometry a circumference of a circle is 360°, and each degree is divided into minutes (small distances) and it happens to be 60, but then multiply 360 by 60 you get 21,600 minutes of Arc, taking that forward 21,600 multiplied by 60 you get 1,296,000 seconds. Bang, we have 1296 appearing again, coincidence?

Geometry Measurements

The geometry is interesting, as a circle is made up of 360°

A square has 360° made up of 4 x 90° angles in each corner

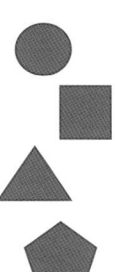

An equilateral triangle has 3 x 60° angles totalling 180°

And a pentagon has 5 internal angles of 108 totalling again 360°

The facts start to reveal themselves as follows:

- The circle is 360° or 36, add the 3 and 6 you get = 9
- The square 4 x 90° = 360 or 36, add the 3 and the 6 you get = 9
- Triangle 3 x 60° = 180 or 18, add the 1 and the 8 you get = 9
- The pentagon is 5 x 108° = 540 or 54, add the 5 and the 4 you get = 9
- 1296000 adds to 9, this is a Tesla number which we will visit in another chapter

Three Dimensions

Let us move into the physical 3 dimensions. Polyhedrons (regular faces) are shapes that have the same area sides that perfectly make a shape of which there are only 5 of these 3-dimensional shapes; tetrahedron, cube, octahedron, dodecahedron, and icosahedron.

There are some interesting numbers with these:

The tetrahedron 4 vertices (or faces) or 4 equilateral triangles, 6 edges with 3 edges on each face, each of its 4 faces has 180°.

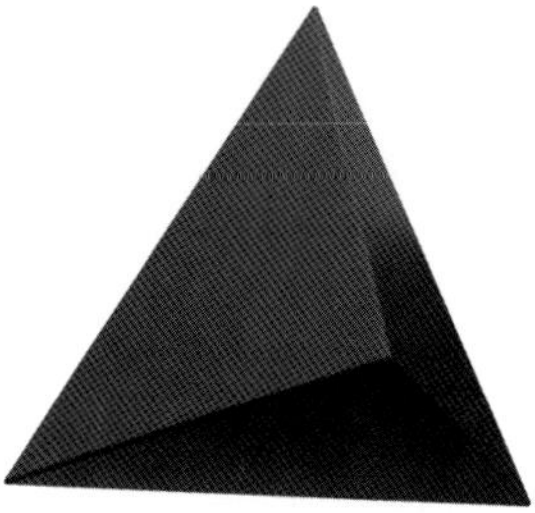

Multiply 4 x 180 and you get 720°. Now consider the total minutes in a 12-hour cycle = 720.

Then there is the cube which holds 6 faces.

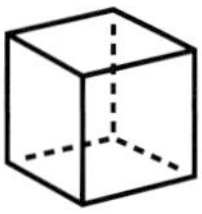

8 vertices, 12 edges, 6 faces each with 4 sides. Each face has 4 x 90° angles totalling 360 but multiply this by its number of sides and something remarkable happens 6 x 360 = 2,160 *space measurement number*.

The third of the 3D shaped polyhedrons is the octahedron or 2, five sided pyramids placed bottom to bottom.

It has 6 vertices, 12 edges, 8 faces with 3 sides and 180°, so 8 x 180 = 1440° and now it is a repeat of a *time measurement number* of 1440, 144 sq. inches, 1,440 minutes in a day so this number is again special in nature.

The kicker is that if the Earth makes one complete spin or rotation this totals 1440 minutes.

The fourth polyhedron is the dodecahedron with 20 vertices, 30 edges which has 12 faces and 5 sides.

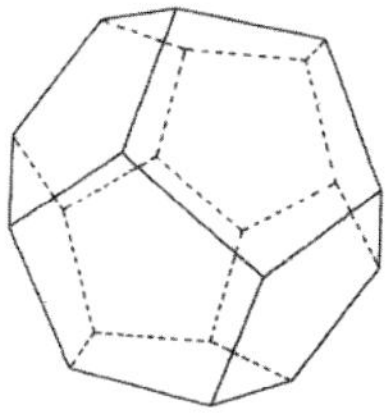

Each of the faces of 12 x 540° is 6480° or 5+4+0 is 9.

As the final basic polyhedron (shapes that have the exact same faces), the icosahedron has 20 faces. All equilateral triangles.

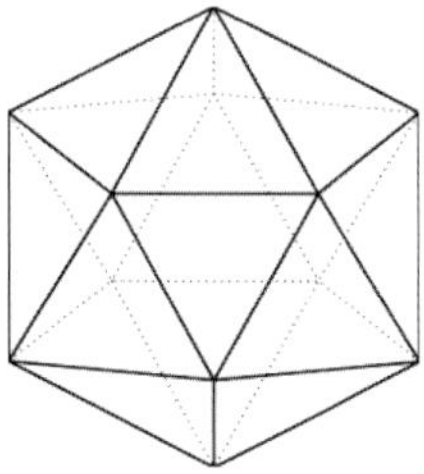

12 vertices, 30 edges, 20 faces each with 3 sides x 180° results in 360° once again 3+6 = 9.

This group is called the platonic solids named by Plato (Greek philosopher 347 BC) and he described these shapes in accordance with Earth's elements the Tetrahedron-Fire, Cube-Earth, Octahedron-Air, Icosahedron-Water, Dodecahedron-Universe.

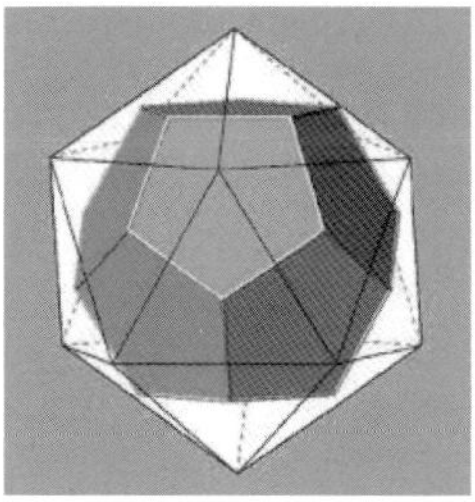

Interestingly the icosahedron fits perfectly inside a dodecahedron with its triangulation peak at perfect centre and meeting the of the face of the latter. Hold these numbers in your head as we deep dive into time.

Astrological Time Measurements

The Earth has its daily motion of one complete spin (daily cycle), its second is its orbit around the Sun (yearly cycle) but there is the third called the processional cycle which is a slow wobble on its own axis like a pendulum rotation. Astrophysicists call it the "G*reat Year",* which is a cycle of 25,920 years. If you take this number and divide it by the 4 seasons for each year you get 6,480 years or 4 durations of 6480 years which is the exact number of the dodecahedron. This 3rd cycle is driven by the Moon which we will explore elsewhere in this book.

Time Measurements

We have solar days consisting of 24 hours, with each hour made up of 60 minutes or a total of 1,440 minutes in a day which is 86,400 seconds. Our 12-hour time cycle is 720 minutes or 43,200 seconds. An interesting fact that 144,000 is referred to Biblically in Revelation 14:1-4 *"And I looked, and behold, a Lamb stood on the mount Sion, and with him a hundred and forty four thousand"*.

Cosmic wheel and the Mayans

I believe ancient cultures knew far more about time than we do. The Mayans (circa 1000 BCE) were of Mesoamerica or modern-day Mexico and Central America. They used a basic unit measurement of time and one day was called *kin*. This period of measurement moved upwards in jumps of 20. They believed that the *great cycle* was 13 tuns (1,872,000 kins). This chronological count often referred to as the long count stretched to 5,128 years with the *great cycle* ending in the winter solstice of 21st December 2012. Much doomsday prediction was made during this year as I remember but in fact it was just the end of a cycle. Looking into the numbers, however, you will now see some striking coincidences.

Cycle	Length	Number of Days
1 B'AK'TUN	20 K'ATUN	144,000 Days
1 K'ATUN	20 TUN	7,200 Days
1 TUN	18 WINAL	360 Days
1 WINAL	20 K'IN	20 Days
1 K'IN	0-1	1 Day

This continues with:

- 1440 is both time and space but add 1+4+4+0 = 9
- 8640 is time but add 8+6+4+0 = 9
- 1728 is space but add 1+7+2+8 = 9
- 21600 is space but again add 2+1+6+0+0 = 9

The oldest and most ancient, recorded civilisation was that of the Sumerians. Their homeland was called Mesopotamia or Sumer, approximately 6000 years ago and was along the floodplains between Tigris and Euphrates rivers in present-day Iraq, Syria, Kuwait and part of Turkey.

Looking at the tablets of the Sumerians which predates our time calendar and according to translations of Zachariah Sitchin (theologist of the Sumerian culture) they had within their recorded history of pre-flood or pre-diluvian God periods with the years in time which they ruled. A striking pattern appears from all the numbers I previously mentioned, and I have listed below for clarity:

Name of the Kings	**Years in which they reigned**	
ALOROS	36,000	3+6 = 9
ALAPOROS	10,800	10+8 = 9
AMELON	46,800	4+6+8 = 18 and 1+8 = 9
AMMENON	43,200	4+3+2 = 9
MEGGALAROS	64,800	6+4+8 = 18 and 1+8 = 9
DAONOS	36,000	3+6 = 9
EUEDOROCHES	64,800	6+4+8 = 9
AMEMPSINOS	36,000	3+6 = 9
OPARTES	28,800	2+8+8 = 18 and 1+8 = 9
XLSUHTROS	64,800	6+4+8 = 18 and 1+8 = 9
Total Years	432,000	4+3+2 = 9

Researching the Sumerian Texts, the earliest recorded stone tablet is from the Taurean Age some 6,480 years on the 25,920-year cycle. Here we have 6480 again. What is striking is the 12 signs of the zodiac when divided by the processional cycle of 25,920 equates to 2,160 years. Again 2160 appears which as above is a geometric sacred number. Each zodiac cycle of 2160 years consists of 30 x 72 Earth cycles.

By now you should be unsettled as we are demonstrating that time, shapes, distance, and the planetary movements are all related by exact numbers. Here are some more coincidences:

- 6,480 years 6+4+8 = 18 and 1+8 = 9
- 2,160 years 2+1+6 = 9
- 72 Earth cycles 7+2 = 9

The numbers are clearly embedded in time as per below:

25,920 seconds	=	432 minutes
432 minutes	=	7.2 hours
12,960 seconds	=	216 minutes
6480 seconds	=	108 minutes
25,920 hours	=	1080 days
25,920 days	=	72 years of 360 days

In their culture, days equate to years, years equate to hours and time relates to shape and all with the number 9.

The Sumerian Texts shows they worked on a cycle of 360 days and due to a cataclysmic event, the Earth suffered a celestial alteration adding 5 days to the yearly cycle. For this reason, they operated on two calendars, one of the years and one of celestial movements and so was born the *signs of the zodiac* and planetary positioning of the Earth in each of the 4 x 6480 years in the complete 25,920 years.

Let us not forget 25,920 years cycle 2+5+9+2 = 18 and 1+8 = 9.

Time, space, and geometry are all connected numerically.

You will never look at a pizza the same again.

Chapter 3 - Numerology and Gematria

The *stopper 1* glanced off these mystery school numbers but I did not drill into their significance. Firstly, we must de-code a few things. This chapter is at the heart of the mystery schools, let us explore the subject.

People often mix up the term's *astronomy* and *astrology,* but they are two separate subjects as is *gematria* and *numerology.*

Numerology

The use of numbers to interpret a person's character or to predict the future. The theory behind numerology is based on the Pythagoras idea that all things can be expressed in numerical terms because they are reducible to numbers which is why I reduced the previous chapter numbers using addition to a base single number. Using an analogous method (comparing two things as the same) in the Greek and Hebrew alphabets (in which each letter also represented a number). Modern numerology attaches a series of digits to an individual's name and date of birth and from these purports to divine the person's true nature, character, and prospects (Encyclopaedia Britannica). Numerology is the use of numbers (and names converted into numbers). The Encyclopaedia continues to explain that its use was used to interpret a person's future *'by a system that assigns particular metaphysical and natural significance to various integer values'* the term can also refer more generally to the study of the "hidden meaning" of numbers.

This pseudoscientific belief is that in the divine a mystical relationship exists between numbers and letters.

English for example A=1, B=2, C=3, D=4, et cetera.
Reversed Ordinal is A=26, B=25, C=24, D=23 etc.

There are many different ordinals such as Jewish, Full Reduction, English Ordinal, Satanic, and others which are in the back of this book.

My name is *"gary patrick Fraughen"* so let us calculate this out in numerology under Full Reduction. 7+1+9+7+7+1+2+9+9+3+2+6+9+1+3+7+8+5+5=101. The mystery schools regard this number as *"one of change, new beginnings"*, and more worryingly *"the end of things"*. I am therefore said to have these traits and an ability to deliver them.

When you think of the number 101 in literature and entertainment you start to see its use. Room 101 in George Orwell's 1984 is a room in which those who went into never came out. In the Terminator movies, Cyberdyne Systems Model 101 was used to time travel to try to destroy and alter humanities past and future. The purveyors of power use this number as an expression of energy and when you get to the coronavirus section of this book you will see the sinister nature when its power is used. This is deep in cabalistic calculations but if you take 1+0+1=2. Now if you look at my name in the English Ordinal you get 7+1+18+25+16+1+20+18+9+3+11+6+18+1+21+7+8+5+14=209, and 2+9, you get 2. My numbers are then considered as 2 and 101.

Calculate your own name out, you may find some interesting numbers. Still to this day the Arabic, Chinese, Hebrew, Japanese cultures use this for match making, decision making and important dates in astronomy or planned events, so do the mystery schools.

Gematria

The substitution of numbers for letters of the Hebrew alphabet, a favourite method of exegesis was used by Medieval Kabbalists to derive mystical insights into sacred writings or to obtain new interpretations of the texts. Some condemned its use as mere toying with numbers, but others considered it a useful tool, especially when difficult or ambiguous texts otherwise failed to yield satisfactory analysis. Genesis 28:12, for example, relates that in a dream Jacob saw a ladder (Hebrew *sullam*) stretching from earth to heaven. Since the numerical value of the word *sullam* is 130 (60+30+40 = 130), the same numerical value of Sinai (60+10+50+10 = 130) exegetes concluded that the Law revealed to Moses on Mount Sinai is "man's means of reaching heaven" (Encyclopaedia Britannica). Words therefore have association by numbers if they add to the same number.

When the original Bible was written there were no actual symbols for numbers as mentioned in chapter 1. Written communication as symbols stood in place of numbers that we use today. Within Revelation, 666 did not exists it was ςςς in the Greek and Hebrew for example.

These letters had in effect a numerical relationship with numbers not as we see them today, but this association continued in time.

Although the term Gematria is Hebrew, it may be derived from the Greek γεωμετρία geōmetriā, "geometry", which was used as a translation of gēmaṭriyā, although some scholars believe it to derive from Greek γραμματεια grammateia "knowledge of writing". The original Greek called it isopsephy from isos-equal psephos-pebble, as the Greeks used pebbles to count so it became a counting term. As a word it abridged making sense of numbers visually. The Dictionary defines it as *a "cabbalistic method of interpreting the Hebrew scriptures by interchanging words whose letters have the same numerical value when added"*. The term Gematria is also argued that its origin could be either of the above, or a mixture of the above. Gnosis (knowledge) is a city in Greece, Knossos but Gematria in Kabbalah is based in geometry and a divine relationship with nature.

The Hebrew word *Chai* (meaning *life*) חַיִּים. This two-letter composition adds to 18 and 1+8 = 9. You will need to turn to the last pages of this book and use it as a reference point when you read the pages relating to the *Gematria codes.*

Unless Gematria is being used in conjunction with nature and geometry then it is not being used to its full potential. It has its important foundations in geometry, angles, and measurements.

Nature's most natural shape, the circle, known as the definition of time – a form of Matrix Decoded: The word matri-x or matrix (whose plural is matrices) is a rectangular array of numbers, symbols, or expressions, arranged in rows and or columns.

English Ordinal
Hebrew and Greek are where gematrias originate but with the English language being heavily influenced by Latin script it too has a valid place. A quite simple correlation I found when I entered these word calculations were hidden numbers. The word GOD is G=7 O=15 and D which is 4, add the numbers together 7+15+4 = 26 which is simple enough. Then take the word Church = 101 in reversed code. Why was I interested, well because 101 is the 26th prime number so the word GOD lays mathematically in CHURCH.

666 is mentioned throughout other chapters from here on but when you consider maths is all around us certain numbers keep appearing.

It is now common knowledge that when the index finger touches the thumb and shown to an audience the remaining fingers demonstrate the number 666, however, on the qwerty keyboard the same index finger is the number placed upon the number 6 before the typing starts.

In biblical terms 666 is the number of the beast but this deepens in mathematical terms. The golden ratio also known as the divine proportion which is Phi = 1.6180339887...that does not seem to be important but stop. If you move to trigonometry and take the following calculations:

- Sin (666°) = Cos ([6x6x6]) it starts to break down (-.80901699437) + (-.80901699437) so now it is getting strange, but add them and boom = - 1.6180339887, which is the complete inverse of the golden ratio or the holiest number in mathematics, just a coincidence?

As I mentioned Rome used to use 6 alphabet letters as a number codex to count as follows, D=500, C=100, L=50, X=10, V=5, I=1. If you add the numeral numbers below 1000 or DCLXVI.

D=500
C=100
L= 50
X= 10
V= 5
I= 1 +
----
6 66

Just as the Romans used letters so did the Greeks and the Hebrews. The Hebrew words Nero Caesar as NRON QSR which adds to 666. Just another coincidence well oddly when we do the same with Greek, we get 666.

222 is another number that will creep up here and there in secret societies, this number is for birth, death and rebirth but look at other words that add to these energy numeri.

- Simple English Gematria
  19+9+13+16+12+5+5+14+7+12+9+19+87+5+13+1+20+18+9+1 = 222
- English-based isopsephia = 222
- Letters as numerals = 222
- To count from A to Z = 222
- Alphabetic numerology = 222
- Alphanumeric system = 222
- The secret of the alphabet = 222
- It is hidden in the alphabet = 222
- Written patterns = 222
- Impossible coincidences = 222
- Numerous parallels = 222
- It's two-two-two = 222
- Twin tower attacks = 222
- A great satans visit Dec 2021 = 222
- Covid 19 vaccines out in 2020 = 222
- Trump to leave office 2021 = 222
- World control fails in 2022 = 222
- Event two zero one 5+22+5+14+20+20+23+15+26+5+18+15+15+14+5 = 222
- Order out of chaos RO = 222

In the Bible the words truth and wisdom are both found in exactly 222 verses of the King James 5th edition.

I will show you some gematria in the next chapter and from then on randomly in others on specific words, which will add another layer and dimension to your read. If numbers are behind the universe, they are without doubt in my opinion behind words, statements, and language we use, and some are hidden with sinister intent.

Chapter 4 - Numbers

Ask, and it shall be given to you; seek and ye shall find; knock, and it shall be opened unto you...Matthew 7:7.

We will explore some of the basic numbers in this chapter. Words we use every day have hidden number codes sleeping within the letters. While the English language is the most diverse written language, as a construct that has taken thousands of years to arrive in its current format the codes are present. Here are some words and numbers.

1

The fear of number 1 is *Henophobia* (from Greek-*hen*) or *unophobia* (from Latin-*unus*) meaning 'one'.

Birthday candles are a ritual prayer for witchcraft. Blowing out a candle is known as candle magik. We are told to make a wish, close your eyes, expel the flame, but not reveal what your demands are. The pagans use to worship 1 lit single candle in honour of Artemis (goddess of the Moon). The Egyptian Pharaohs used to celebrate their coronations each year with a sugared offering, placed upon a waxed effigy with a lit candle placed on top. These pagan celebrations became spiritualist (not spiritual). In the West, the Germans pioneered the *kinderfest* a celebration of a first-year old's success in staying alive. Most children died during this critical 1st year of life and the candle started to become widely used, with frosting on a cake to give the sugar creation longevity but keeping out oxygen. In 1893 Patty and Mildred J. Hill penned the song '*Happy Birthday*' (originally called '*Good Morning to All*') and up until 2016 the song was under copywrite to Warner/Chappell Music. They collected a staggering $14 million from movie and music rights, only to fail in its continuance following a US Supreme Court ruling in 2016. It is now in the public domain.

- 1st March - Salem witch hunt begins
- 1st March - Roman empire's New Year's Day
- 1st March - US test nuclear weapons in Bikini Atoll
- 1st May - Act of Union between England and Scotland, creating the United Kingdom of Great Britain
- 1st May - Bavarian Illuminati was formed
- 1st May - Planet Pluto is officially named

2

The fear of number 2 is *Dyophobia* (from Greek-*dyo*, meaning 'two'), also known as *diphobia* or *duophobia* or the fear of pairs.

- Matter and cell doubling
- Balance things being even
- Pinched up energy, the binding of two cells
- 2 pillars Boaz and Jachin
- Janus worship - 2 faced
- Diplomacy, two sides to a story, middle road story, contrast, harmony
- Co-operation

- Hope and fear, win and lose
- It is the number of dualities, the start of something. 1+2 is something that is growing. From subatomic particles to the vastness of the cosmos expansion, starts with 1+2. From the subconscious mind to the collective unconscious universe. In terms of linear travel from point A to point B, mathematics is 1 towards 2
- Is equal and opposite, think of this number in terms of a balance or a scale of measure which is equal in its 2 parts
- The number for the next step, such as creating, building, thinking, and planning. 2 is the start of a process. A bridge has a start and finish
- Is Janus worship explored elsewhere in this book

Parents often ask me; how would you explain politics to a child, and I simply use the 2-cow number method:

Socialism - You have 2 cows. You must give 1 cow to your neighbour.
Communism - You have 2 cows. You must give 2 cows to the government, and they may give you some milk.
Nazism - You have 2 cows. The government shoots you and takes 2 cows.
Fascism - You have 2 cows. You must give all the milk to the government and the government sells it.
Capitalism - You have 2 cows. Then you buy a bull to make more cows.
Anarchism - You have 2 cows. You fight to keep 2 cows shoot the government agent and steal another cow.

3

The fear of number 3 is *Triskaphobia*, also known as *triphobia* and *triophobia* (from Greek- *tria* or 'three').

In secret societies this number reflects the day of worship of Nimrod or Baal.

- December 25th is in planetary terms the rebirth of the Sun with our Seasons. This day is 3 days after the Winter Solstice (the shortest day and the longest night) in Earth's calendar
- 'Chant of three' adds to 3+8+1+14+20+15+6+20+8+18+5+5 = 123 and 1+2+3 = 6
- There is even a song called three steps to heaven
- Take any number and multiply it by 3. Then, take the digits of that new number and add them all together. Whatever number that equals it will always be divisible by 3, no matter what number you started with. For example:
    - 3 x 4 = 12 and 1+2 = 3
- 3 little pigs
- 3 primary colours - red, blue, green
- 3 bears
- 3 Billy Goats Gruff
- 3 blind mice
- Father, son, and holy spirit
- Ancient Babylon 3 primary gods Anu, Bel (Baal) and Ea
- Heaven, Earth, and the abyss

- 3 witches of the Scottish play (1606)
- 3 is the smallest dimension of magic square which sums to 15 discussed elsewhere in this book
- 3 months, 3 weeks and 3 days is a pig's gestation period
- 3 demonstrates human capability of thought, word, and deed
- 3 bones in the human ear - hammer, anvil, stirrup
- 3 demonstrated propositions - subject, predicate and copular
- 3 demonstrates matter - mineral, vegetable, and animal
- 3 gifts of grace - faith, hope and love
- 3 elements of the existence of consciousness - body, mind, and spirit
- 3 wise men
- 3 feet in a yard
- 3 tenses in the English language - past, present, future
- 3 gifts to baby Jesus - gold, frankincense, and myrrh
- Rock, paper, scissors
- 3 monkeys - hear, see, and speak no evil
- Triskelion - goddess, 3 legs of celts
- <u>Time</u> - past (yesterday is the past), present (today is a gift, that is why it is called a present) and future (tomorrow is the future) = 3
- <u>Space</u> - height, width, and depth = 3
- <u>Matter</u> - solid, liquid and gas = 3
- Love, peace, and unity
- 3 Christian positions to worship God - standing, kneeling, and bowed
- 3 Laws of nature - birth, life, and death
- 3 AM the witching hour, said to be when dark energy is at its most powerful
- 3 main particles - protons, neutrons, electrons
- 3 Laws of water - ice, liquid, vapour
- 3 versions of history - one side, the opposing side, and the truth
- 3 Fundamental forces - Strong Nuclear, Weak Nuclear and Electromagnetic
- Atomic number 3 - Lithium deuteride is the explosive material of the hydrogen bomb and may eventually be the fuel of controlled fusion reactors. There is great power in three

Phrases that are repeated 3 times in the Bible:

- *"Before the foundation of the world"*
- John 17:24 "*Thou lovedst Me before*", etc
- Ephesians 1:4 "*Chosen us In Him before*", etc
- Peter 1:20 "*The blood of Christ foreordained before*", etc. (when it speaks of this blood as "shed" it is from the foundation)

Words repeated 3 times in the Bible:

- (ah-dar), glorious, Exodus 15:6,11; Isaiah 42:21
- (aph-see), beside me, Isaiah 47:8,10; Zephaniah 2:15
- (g'moo-lah), recompense, 2 Samuel 19:36; Isaiah 59:18; Jeremiah 51:56
- (nah-cheh), lame, Samuel 4:4, 9:3; contrite, Isaiah 56:2
- (at-teek), ancient, Daniel 7:9,13,22

- (r'phoo-oth), medicines, Jeremiah 30:13, 46:11; Ezekiel 30:21
- (abba), father, Mark 14:36; Romans 8:15; Galatians 4:6
- (haireomai), to choose, Philippians 1:22; Thesalonians 2:13; Hebrews 11:25
- (apokruphos), hid, Luke 8:17; Colossians 2:3; kept secret, Mark 4:22
- (apophthengomai), speak forth, Acts 26:25; utterance, Acts 2:4; say, Acts 2:14
- (acheiropoieetos), made without hands, Mark 14:58; Corinthians 5:1; Colossians 2:11
- (euodia), sweet savour, Corinthians 2:15; Ephesians 5:2; Philippians 4:18
- (kateuthuno), guide or direct, Luke 1:70; Thessalonians 3:11; Thessalonians 3:5
- (morphee), form, Mark 16:12; Philippians 2:6,7

Bees

- In 3 days, the egg of the queen is hatched
- It is fed for 9 days (3x3)
- It reaches maturity in 15 days (5x3)
- The worker pupa reaches maturity in 21 days (7x3)
- It is at work 3 days after leaving its cell
- The drone matures in 24 days (8x3)
- The bee is composed of 3 sections - head and 2 stomachs
- The 2 eyes are made up of about 3,000 small eyes, each (like the cells of the comb) having 6 sides (2x3)
- Underneath the body are 6 (2x3) wax scales with which the comb is made
- It has 6 (2x3) legs, and each leg is composed of 3 sections
- The foot is formed of 3 triangular sections
- The antennae consist of 9 (3x3) sections
- The sting has 9 (3x3) barbs on each side

The alphanumeric sum of NORTH, SOUTH, EAST, WEST is 270, a number which reduces to 9 and is, of course, three 3's. It so happens that the only combination of 2 directions that sums to a 3-number is NORTH and EAST:

(NORTH)75 + (EAST)45 = 120 = 1+2+0 = 3

4

The fear of number 4 is *Tetraphobia* (from Ancient Greek-*tetrás* or 'four').

There is only one number spelled with the same number of letters as itself. Can you guess which one? No? Well, it is 4. Oh, and the number 4 on a calculator is made up of four light bars. The number 4 is viewed with superstition and distrust in much of East Asia. The reason may be because the word for 'four' sounds like 'death' in several Asian languages, including Chinese, Japanese, and Korean. In Ireland, however, a four-leaf clover (shamrock) is considered one of the luckiest things you can find.

In mediaeval times there were thought to be 4 humours (phlegm, blood, choler, and black bile (melancholy)), hence the adjectives phlegmatic, sanguine, choleric, and melancholic. The body was bled at various places to bring these humours into balance.

- 4 elements of metaphysics - being, essence, virtue, action
- 4 elements - earth, fire, water, air
- 4 seasons - spring, summer, autumn, winter
- 4 years between each Olympic Games
- 4 points of the compass (also for information) - North, East, West, and South (news)
- 4 phases of the Moon - also known as Lunar Phases: First Quarter, Full Moon, Last Quarter and New Moon. An additional 4 intermediate phases make up the combined 8 phases that comprise the Phases of the Moon in the following sequential order: New Moon, Waxing Crescent, First Quarter, Waxing Gibbous, Full Moon, Waning Gibbous, Last Quarter and Waning Crescent
- 4 cycles of life - birth, growth, experience, death
- 4 periods of the day - sunrise, midday, sunset, midnight
- 4 Pillars of Destiny from the Chinese Han Dynasty - year, month, day, hour of birth
- 4 types of animals - creeping, flying, running, swimming
- 4 gods to the Native American Indian - superior, ally, subordinate, spirit
- 4 ages to a human - infant, child, mature, elderly
- 4 Horsemen of the Apocalypse - white, red, black, and pale horses
- 4 punishments of God
- 4 Masonic Virtues - temperance, fortitude, prudence, justice
- 4 Evangelists and Gospels - Matthew, Mark, Luke, and John
- The word 'forth' appears 888 times in the King James Bible, more interestingly in Psalm 88:8 ends with the word 'forth'
- The Hindus believe the 4th world is a chain of spheres
- Jewish Genesis states the Earth was formed on the 4th day
- 4 quarters of the horizon
- 4 displays of man's evil tendencies - inclination, thoughts, words, and evil actions
- 4 humans who are little better than dead in the Talmud - the blind, leper, pauper, and those who have no sons
- 4 Canopic Jars were used in mummification of the Pharaohs containing the 4-GENII
- 4 Great councils of the Christian Church - Nicaea, Constantinople, Ephesus, and Chalcedon
- 4 periods of time - o'clock, quarter past, half past, quarter to
- 4 in the mystery schools epitomise - 4 sides, 4 corners, 4 corners and the square
- 4 emotional states of the body - physical, emotional, sensation and spiritual
- 4 states of matter - gas, liquid, solid and plasma
- 4 forces of known nature that bind matter - weak, strong, gravity and magnetic
- 4 philosophies of Buddhism - compassion, affection, love, and impartiality
- 4 *Lokas* of Hinduism - Satya-loka (Brahma-loka), Tapa-loka, Jana-loka, Mahar-loka

In the Arabian nights, the 9 features of a woman based on 4:

- 4 black - hair, eyebrows, eyelashes, and eyes
- 4 white - skin, white of eyes, teeth, and legs
- 4 red - tongue, lips, cheeks, and gums
- 4 long - back, fingers, arms, and legs
- 4 wide - forehead, eyes, seat, lips
- 4 fine - eyebrows, nose, lips, and fingers

- 4 thick - buttocks, thighs, calves, and knees
- 4 small - breasts, ears, hands, and feet

4 types of teachers in the mystery schools:
1. he who learns and then will not teach
2. he who wants to teach and does not learn
3. he who learns and then teaches
4. he who listens and will not learn and cannot teach

5

Fear of number 5 is *Pentaphobia* (from Greek-*penté*) or *quintaphobia* (from Latin-*quinque*) meaning 'five'.

It is the number least feared of all numbers by numerophobic people which is strange when you think about the pentagram.

How did the 5-day week come about? Do you know?
Imagine you worked at the Ford Motor Company on 25th September 1926, suddenly you were told that your working week would fall from 6 days to 5 days and 48 to 40 hours. Added to this the company raised the male employees pay to $5 a day ($116 day in today's money) which was a doubling of the standard pay at that time. Productivity skyrocketed; retention was never an issue as engineers from all over the world wanted to work for them. 8% of employees could now afford to purchase a car they were manufacturing, leisure time increased which meant more clothing was bought, pocket watch sales had waiting lists and other manufacturers had to follow suit. Employees started to buy, then build new homes and the capitalist system expanded at great speed across America. Henry Ford was quoted as saying "*The harder we crowd business for time, the more efficient it becomes. The more well-paid leisure workers get, the greater become their wants. These wants soon become needs. Well managed business pays high wages and sells at low prices. Its workers have the leisure to enjoy life and the wherewithal with which to finance that enjoyment*".

- 5 Pillars of Islam - Declaration of Shahadah (Faith), Salat (Prayer), Zakat (Alms), Sawn or Ramadan (Fasting), Hajj (Pilgrimage to Mecca)
- The Prayer (Salat) is observed five times a day; Farj (dawn), Zuhr (noon), Asr (afternoon), Maghrib (sunset), Isha (dusk)
- Azan (also called Adhan) call to prayer 5 times a day in the Muslim faith traditionally from the minaret
- 5 architectural columns - Tuscan, Doric, Ionic, Corinthian and Composite
- 5 categories of Islamic Law - obligatory, recommended, permitted, dis-liked, forbidden
- 5 law giving religious prophets - Noah, Abraham, Moses, Jesus, Muhammad
- 5 modes of Theology regarding the Conception of God - Pantheism, Polytheism, Dualism, Unitarianism and Trinitarianism
- David used 5 stones to fight Goliath
- 5 principal parts to Solomon's Temple - the pair of columns at the entrance, the forecourt, the Outer Sanctum, the Holy of Holies, and the side chamber
- 5 things permitted to be killed on the Sabbath - the fly in Egypt, the wasp in Nineveh, the scorpion of Hadabia, the serpent of Israel, and the mad dog anywhere

- 5th Elements are earth, air, water, and fire. The fifth element refers to what was known as the aether, a special unknown substance that permeated the celestial sphere and was purer than any of the four terrestrial elements
- Vitruvian man by Leonardo da Vinci, 1 head 2 arms, 2 legs
- 5 atmospheric layers of the Earth - exosphere, troposphere, stratosphere, mesosphere, thermosphere
- 5-pointed star or pentagram
- 5 duties for Christians – holy festivals, fasting, public worship, receive sacraments and adhere to customs of the church
- 5 senses
- 5 fingers including the thumb
- 5 toes
- 5 the planet Mercury from Roman mythology - the God of communication
- 5 in the Hermetic Tarot Deck cards represents an indicative identification of a problem
- 5 foundations of mindfulness - contemplation, body, feelings, mind, objects
- 5 virtues of mankind to live in harmony - generosity, fellowship, purity, courtesy and mercy
- 5 triangles of Hindu symbolism or five points of connectivity - knowledge, senses, organs, breathing exercise and pure consciousness

5 plains of the mystery schools:

1. The great Plain of Chemistry (atoms to cells)
2. The great Plain of Biology (cells to animals)
3. The great Plain of Psychology (animals to humans)
4. The great Plain of Metaphysics (humans to angelic)
5. The next Plain of Existence (The unknown)

These 5 phases form a schema of the Tree of Life:

1. Source or Seed - The Ain Soph Aur or Adam Kadmon
2. Root - World of Emanation of Atziluth
3. Tree - World of Creation or Briyah
4. Branch - World of Formation or Yetzirah
5. Fruit - World of Action or Assiah

6

The fear of number 6 is called *Hexaphobia* (from Greek-*hex*) or *sexaphobia* (from Latin-*sex*) meaning 'six'.

The first 3 degrees in mystery schools have 6 acts of completion in each degree and at the penultimate 3rd degree means there has been 66 and 6 stages to this low-level illumination. *The Secret Doctrine, the Synthesis of Science, Religion and Philosophy*, a book originally published as two volumes in 1888 written by Helena Blavatsky (according to Wiki). The first volume is named *Cosmogenesis*, the second *Anthropogenesis*. It was an influential example of the revival of interest in esoteric and occult ideas in the modern age, in particular because of its claim to reconcile ancient eastern wisdom with modern science. The book has been criticised for promoting pseudoscientific concepts and for borrowing those from other systems. Unfortunately, the factual basis for Blavatsky's book is non-existent. She claimed to have received her information during trances in which the Masters of Mahatmas of Tibet communicated with her and allowed her to read from the ancient *Book of Dzyan*. The *Book of Dzyan* was supposedly composed in Atlantis using the lost language of Senzar but the difficulty is that no scholar of ancient languages in the 1880s or since has encountered the slightest passing reference to the *Book of Dzyan* or the Senzar language.

- 6 weeks - 10 seconds, that is 10 × 9 × 8 × 7 × 6 × 5 × 4 × 3 × 2 × 1 = 3,628,800 seconds = 60,480 minutes = 1,008 hours = 42 days = 6 weeks. 6 is the number of Time itself
- 6 days of creation
- 6th day in Genesis when man was created
- 6th day of the week, is when Jesus died upon the cross
- 6 is the atomic number of Carbon, with 6 electrons 6 protons and 6 neutrons
- 6 - The occult practice of a *séance* (French 'sitting') involves the balance of 3+3 totalling 6 participants
- 6 directions - up, down, forward, back, right, left

7

The fear of number 7 is called *Heptaphobia* (from Greek-*Hepta*, meaning 'seven').

Rosicrucians and 7 planets:

Si	Moon	Silver
Ut	Mercury	Quicksilver
Re	Venus	Copper
Mi	Sun	Gold
Fa	Mars	Iron
Sol	Jupiter	Tin
La	Saturn	Lead

The Rosicrucian's had a relative value of 7 in the musical scale and in the ancient planetary formulas as above. These secret harmonical sounds, drawn into poems, songs and sung at ceremonies which are still used today. These phonics are important even within church hymns.

Within some of Shakespeare's work is the periodic reasoning of 7, for example the child was not named before its 7th day of the birth and it was not thought to be part of the family up to the 7th month. The teeth spring out of the child's mouth on the 7th month and were renewed in the 7th year. At the third 7th (21) the infancy ends, and adulthood commences, and we become legally competent and responsible for our acts. 4×7 = 28 a man is in full possession of his strength and at 5×7 = 35 he is fit for the business world. At 6×7 = 42 he is wise and at 7×7 = 49 he decays. At 8×7 = 56 he is in his first elimacteric and by 9×7 or 63 he is in his grand elimacteric or his final years of danger. And 10×7 or 3 score years and 10 is pronounced to have succeeded the maximum natural period of human life. To Rosicrucian's the number 7 is a growth number.

Historically the rule of 7 regarding babies were as follows: 7 days a child will live, 7 days the cord falls off, 2x7 days the eyes will follow light, 3×7 days they turn their head, 7 months the teeth grow, 2x7 months they are able to sit, 3x7 months begins to talk, 4×7 months able to walk properly.

We believe from the Old Testament that animals were led into the Ark 2 by 2. If you read the Bible, you will realise that only unclean animals were in pairs, clean beasts were permitted in 7 x 7. Genesis 6:20 and Genesis 7: 2–3.

- 7 exists in the ancient systems. There were for instance 7 ancient planets, 7 Pleiades, 7 Hyades, 7 alters burned continually before the God Mithras
- The Arabians had 7 holy temples
- Hindus assumed the world to be enclosed within 7 peninsulas
- The Goths had 7 deities - the Sun, the Moon, Tuisco, Woden, Thor, Friga, and Seatur whose names derive the days of the week.
- In Persian aspirant candidates met 7 obstructions called the road of the 7 stages
- Baal worship - 7 children were used in sacrifice
- The Hebrews required 7 witnesses for union
- The sabbath was the 7th day
- Noah received 7 days' notice of the deluge and is a term still used by modern courts in issuing notices
- 7 persons accompanied Noah into the Ark and it rested on Mount Ararat on the 7th month, the interval between despatching the dove defined land was done on each 7th day
- The walls of Jericho were encompassed for 7 days by 7 priests bearing 7 rams' horns
- Solomon's temple took 7 years to build which was dedicated in the 7th month, the festival lasted 7 days and the tabernacle consists of 7 branches
- The Tower of Babel was said to have been elevated 7 monolithic stories before destruction
- 7 Masons as a minimum required open a lodge
- 7 dwarves
- 7-year cycles in hair growth
- 7-year cycle in bone growth
- 7 brides for 7 brothers
- 7-year itch
- 7th son of the 7th son
- Lucky number 7
- 7 hills of Constantinople
- 7 distinct notes in a musical scale
- 7 colours in the rainbow
- 7 deadly sins - pride, avarice, envy, wrath, lust, gluttony, sloth
- 7 rules in the Hippocratic (*Hippocrates*) oath of medicine, (*Primum-Non-Nocere*) the Latin meaning *'do no harm'* chant of three
- Samson told Delilah to bind him with 7 green withes and one of the 7 locks of his hair
- Israel was oppressed at the hands of Midian for 7 years
- 7 is unlucky in Vietnam
- 7 is the number of the 6 faced dice, explained in the games chapter of this book
- 7 motions of the hand - up, down, left, right, before, behind and circular
- 7 parts of the body - head, torso, left arm, left leg, right arm, right leg, sex organ
- 7 candles in a Menorah

- Balaam told Balak to build 7 alters
- Jacob worked for Laban for 7 years to obtain Rachel's hand in marriage
- Job had 7 sons and 3 daughters making the perfect 10, he also had 7000 sheep and 3000 camels (here we see the number 10 again). His friend sat down with him for 7 days and 7 nights and his friends were ordered to sacrifice 7 bullocks and 7 rams. He also lived 140 years or 7×10 = 70x2
- The interpretation of the dream of King Pharaoh of 7 years of abundance in the land of the Egypt followed by 7 years of famine
- The walls of Jericho fell when the 7 priests and 7 trumpets were used against the city on 7 successive days
- 7 angels in the apocalypse poured out 7 plagues from 7 vials of wrath
- 7 celestial bodies (visible to the human eye) - Sun, Moon, Mars, Mercury, Jupiter, Saturn, Venus
- 7 Old Wonders of the World – Great Pyramid of Giza, Hanging Gardens of Babylon, Colossus of Rhodes, Statue of Zeus, Pharos (Lighthouse) of Alexandria, Mausoleum at Halicarnassus, Temple of Artemis
- 7 New Wonders of the World - Taj Mahal, The Colosseum, Chichén Itzá, Machu Picchu, Christ the Redeemer, Petra, The Great Wall of China
- 7 layers of Islam - 7 earths, 7 skies, 7 heavens, 7 hells, 7 doors to hell, 7 hells itself
- 7 sins of Islam
- 7 sins in Christianity
- 7 or seven as a word in transcendental
- 7 units of measurement in Physics
- 7 directions in flight - left, right, up, down, forward, back, centre
- 7 days of Sukkot
- 7 chakras Hinduism
- 7 is the number of perfections in The Book with Seven Seals and in the Bible St John and the apocalypse

7 classes of Angels:

1. Ishim
2. Arelim
3. Chashmalim
4. Melakim
5. Auphanim
6. Serpahim
7. Cherubim

7 fires that changed the world:

1. The Great Fire of the Library of Alexandria
2. The Great Fire of Rome
3. The Great Fire of London
4. The Great Fire of New York
5. The Great Fire in Chicago
6. Triangle Shirtwaist Factory Fire
7. The Reichstag Fire

The Greek 7 sacred vowels:

1. A - Alpha (Sun)
2. E - Epsilon (Mars)
3. H - Eta (Jupiter)
4. I - Iota (Saturn)
5. O - Omicron (Venus)
6. Y - Upsilon (Moon)
7. X - Omega (Mercury)

The Goths had 7 deities - the Sun, the Moon, Tuisco, Woden, Thor, Friga, and Seatur whose names derive the days of the week.

Monday

Moons day from Moon worship, like the German word Montag and the French word *Lundi* derived from the word Moon.

Tuesday

Tiw's-day was the day of the god of war and combat. Tiu or Norse Tyr was known as the sky god. The French word is Mardi, which is so named after Mars.

Wednesday

Woden's day was the chief Norse god Woden and in French *Mercredi* was after the Roman god Mercury.

Thursday

Thor's-day the god of thunder and lightning and in German *Donnerstag* is donner, 'thunder'. Also named after Jupiter.

Friday

Frige's-day who represented love and in German *Freitag* the Roman goddess of love was Venus.

Saturday

Saturnsday the Roman god of agriculture or Saturn's day.

Sunday

The main Christian religious day but in pagan times it was named after the Sun as sun-worship, which was the first day of the week, also from the German *Sonntag.*

In oriental terms

Sun day, Moon day, fire day, water day, wood day, metal or gold day, earth day.

In Hindi

Ravivaar surya-Sun day, Somavaar-Moon day, Mangalvaar-Mars day, Budhvaar-Mercury day, Brihaspativaar-Jupiter day, Shukravaar-Venus day, Shanivaar-Saturn day.

8

Fear of number 8 is Octophobia.

- The Muslim faith have 7 hells but 8 paradises
- 8 petals of a Lotus flower in Buddhism are considered lucky
- 8 evacuations of the human body - tears, mucus, saliva, semen, menstruation, perspiration, and 2 excretions
- 8 people were aboard the Ark

9

Fear of number 9 is *Enneaphobia* (from the Greak-*ennea*) or (from Latin-*novem*) meaning 'nine'.

9 plus any one digit and the 9 always disappears:

- 9+1 = 10 and 10 or 1+0 = 1
- 9+2 = 11 and 11 or 1+1 = 2
- 9+3 = 12 and 12 or 1+2 = 3
- 9+4 = 13 and 13 or 1+3 = 4
- Etc

9 and 3 in Freemasonry is avoided openly. These fraternities only use even numbers as they were thought of as special, ever heard the term *'he is odd'*, or *'he is an oddball'*. It is an expression of being different from the crowd.

Now that I have explained that in mathematics 9 does not exist, let me show you that there is also something very strange about this number. If you multiply a number by 9 (any number), add all the digits from the result you will find that the new number will always add up to 9. Here are some examples:

8×9 = 72, 7+2 = 9
4×9 = 36, 3+6 = 9
Try 666×9 = 5994, 5+9+9+4 = 27 and finally 2+7 = 9

You can try with any multiple number but remember in mathematics 9 does not exist and it is the reason it is called the *hidden magic number.*

The average number of human breaths per hour is 1080, 1+8 = 9. As mentioned elsewhere in this book 1080 miles is the radius of the Moon. The Moon has always been in direct correlation with the colour silver whose atomic weight is 108, 1+8 = 9.

- The average number of heartbeats of an adult male is 72, 7+2 = 9
- 186282 the maximum speed of light, miles per second 1+8+6+2+8+2 = 27 and 2+7 = 9
- 144000 children of Israel were on Mount Zion, 1+4+4 = 9
- 666 is hiding it, 6x6x6 = 216, 2+1+6 = 9
- Circle divided into 4 hides the fact that each has 90°x4 = 360, 3+6 = 9
- Circle divided into 8 hides the fact that each has 45° and 4+5 = 9

Take another random number that contains 9, for example 719. Add 7+1+9 = 17 then 1+7 = 8 but if we left the 9 out of the original calculation it would be 7+1 which = 8. With more than three numbers one of which being a 9, remove the it and it speeds up the calculation.

In mathematical terms 9 = 0, you have just never seen it.

- 9 realms in Yggdrasil in the ancient Norse mythology with 9 worlds of Asgard which is the realm of the gods
- 9 months for human gestation
- 9 months for lobster's gestation

10

Fear of number 10 is *Decaphobia* (from the Greek-*deka* meaning 'ten').

As is the case for any base in its system, 10 is the first two-digit number in decimal and thus the lowest number where the position of a numeral affects its value.

The metric system is based on the number 10, as all units are divisible by 10. A gram is one hundred micrograms, while a kilogram is a thousand grams. Similarly, a meter is one hundred centimetres, and a kilometre is a thousand meters.

Add up the first 3 prime numbers (2, 3, and 5), you get the number 10. Prime numbers are special numbers in mathematics that can only be divided by themselves and the number 1. Most counting systems around the world use the base-ten numeral system. This means that we count in blocks of ten, such as 10, 20, 30, and so on. Because of this, any number which has a 0 in can always be divided by 10.

The most common paper sizing system uses the A-series, which has 10 sizes. Starting from A0, which is one square meter (10.76 square feet), each following size is half as big as the previous size. This makes A10 the smallest size in this system, measuring just 0.87 inches (22 mm) wide by 1.46 inches (37 mm) long, which is about the same size as a standard postage stamp.

The number 10 is connected to the word *decimate*. These days 'decimate' means a similar thing to destroy, but this was not always the case. The word's history goes back to Roman times, where it meant killing 1 in every 10 people as punishment to rebellious groups within Rome's legions or even to rebellious cities.

- 10 Commandments is the Decalogue
- 10 plagues of Egypt
- According to the Bible, there were 10 generations between Adam and Noah and 10 generations between Noah and Abraham
- 10 seconds determines a knockout in boxing
- 10 acres in a square furlong
- 1/10th of icebergs is under water
- 10% of the world is left-handed
- If you write the number 10 twice as 1010, you get the binary representation of 10

Yin and yang are based on 10:

1	Negative	Positive
2	Feminine	Masculine
3	Passive	Active
4	Intuitive	Logical
5	Dark	Light
6	Night	Day
7	Cold	Hot
8	Soft	Hard
9	Slow	Fast
10	Moon	Sun

Pythagoras and his followers considered 10 the most sacred number, as 10 = 1+2+3+4, which represented existence (1), creation (2), life (3) and the 4 elements, earth, air, fire, and water (4). The mystery schools also consider this a sacred number for the same reason. The Pythagoras followers never had a gathering in groups of more than 10.

11

The fear of number 11 is *Hendecaphobia* (from Greek-*hendeka)* also known as *Undecaphobia* (from Latin-*undecem*) or *Psychicekaidelokaphobia* meaning 'eleven'.

- 11 in a football team
- 11 in a cricket team
- 11 is in a hockey team
- 11 in an American football team
- Elevenses is time to have a break
- 11 years in a sunspot cycle
- 11 Disciples and Judas Iscariot was the 12th who deceived Jesus
- 11 km per second is necessary to escape Earth's gravity
- 11 km is the deepest part of the ocean floor within the Mariana Trench
- 11 ounces is the average weight of a male human heart
- 11 inches is the length of a rugby ball
- 11th hour of the 11th day of the 11th month was to the end of the 1st World War called Armistice Day. Poppy - Traditionally badges are born on the left of a man's lapel, the leaf is always considered important and to be pointing at 11 o'clock
- Apollo 11 was said to be the first lunar landing module
- Attack (FR) = 11
- Black (FR) = 11
- 11 from Old English *endleofan*, meaning [ten and] one left [over]
- K = 11 in numerology and KKK = 11;11;11 so 11+11+11 = 33
- 911 as a date is infamous but 9+1+1 = 11 is a Freemason master number and it means Sin, Transgression and Peril
- Janus (FR) = 11, this word we will explore in another section of this book
- Date of The New York Attack: 9/11 - 9 + 1 + 1 = 11
- September 11th is the 254 day of the year: 2 + 5 + 4 = 11
- After September 11th, there are 111 days left until the end of the year

- Twin Towers, standing side by side, look like the number 11
- The first plane to hit the tower was American Airlines Flight 11
- The State of New York was the 11th State to enter the Union
- New York City consists of 11 letters
- Afghanistan consists of 11 letters
- 'The Pentagon' consists of 11 letters
- Ramzi Yousef, who was convicted of the first WTC bombing, contains 11 letters
- Flight 11, had 92 passengers: 9 + 2 = 11

12

Fear of number 12 is *Dodecaphobia* (from the Greek-*dodeka*) meaning 'twelve'.

- 12 constellations in the zodiac
- 12 hours in a day
- 12 hours in the night
- 12 months in a year
- 12 cranial nerves in the human body
- 12 disciples of Jesus
- 12 inches in a foot
- 12 ordeals of Gilgamesh
- 12 labours of Hercules
- 12 nobles to the Vedic Kings
- 12 names of Odin in Old Asgard
- Some sources say 12 Knights to King Arthur's round table
- 12 Olympians
- 12 oars on the ship of Ra
- 12 Tablets of Law in Rome
- 12 apostles
- 12 days of Christmas
- 12 astronauts are said to have walked on the Moon in the late 1960s and 70s
- 12 permutations of the Tetragrammaton
- 12 recorded appearances of Jesus after his death
- 12 brothers of Joseph
- 12 Judges of Israel
- Testaments of the 12 Patriarchs
- 12 Old Testament Prophets
- 12 Tribes of Ishmael
- 12-step programs
- 12 jurors
- 12 notes in an octave
- 12 in a dozen or dozen eggs
- 12 years of childhood before a you become a teenager
- 12 years girls celebrate becoming Bat Mitzvahs
- Himmler had 12 "Teutonic Knights" of the SS
- A crown of 12 stars in the book of Revelation

- The New Jerusalem has 12 gates
- Even the pagans respected this number for their work in mythology with 12 superior and 12 inferior gods
- There are 12 forms of ceremony within each Freemason degree called the 12 original points: 1 the opening, 2 the preparation of candidate, 3 the report, 4 the entrance of the candidate, 5 the prayer, 6 the circumambulation, 7 advancing to the altar, 8 the obligation, 9 the entrusting, 10 the investiture, 11 the ceremony of the NE, 12 the closing of the lodge
- 12 tribes of Israel - Rueben, Simeon, Ephraim, Judah, Zebulun, Issachar, Dan, Gad, Asher, Naphtali, Manasseh, and Benjamin
- Interestingly in the Bible the fallen angels in mockery and imitation of the gods of the original 12 tribes he blessed 12 specific bloodlines whose names have changed over time, these have become the Astor, Bundy, Collins, DuPont, Freeman, Kennedy, Li, Onassis, Rockefeller, Rothchild, Russell, Van Duyn, but there is another who carried out their work called the Merovingian. Ishmaeli – alchemists assassins' techniques of occult practice specialists
- 12th sign of the Zodiac - Pisces
- Chinese zodiac – rat, ox, tiger, rabbit, dragon, snake, horse, boat, monkey, rooster, dog, pig

13

The fear of this number is called *Triskaidekaphobia.*

- 13 Colonies (known as the Thirteen British Colonies, the Thirteen American Colonies or United Colonies)
- 13 witches in a coven
- NHS admissions increase on Friday 13th as per the Medical Journal 1993
- 13 people at the last supper
- 13th psalm in the Bible is about wickedness
- MS-13 as commonly known for Mara Salvatrucha, a criminal gang in Los Angeles
- 13 weeks exist between Equinoxes and Solstices with the Moon travelling on average 13° a day between each
- 13th letter in the English alphabet results in M which in ancient Hebrew is spelt *Mem* translating from the root word of water and in the same context in Egypt the word for water is move the divine mother associated with water (MOO – n) as all life begins in water
- Traditionally the hanging gallows had 12 steps though it was foretold that there were 12 steps up and 1 step down making it 13 (quite sinister thought)
- The general menstrual cycle of 28 days women generally has 13 cycles in a year
- The devil's luck - is often said that people who are evil or do immoral actions during their lifetime, seem to always be the luckiest and get away with the most heinous crimes or dodge consequences and this is known as the 'devils 13'
- In American Corporate Law, you can file for bankruptcy with all the financial chaos that goes with it is called filing for 'Chapter 13'
- 13 an unlucky number in the West
- Mecca - For those in the West who are unaware of Muslim beliefs, the one thing that Muslins must undertake during their lifetime is the 'hajj'. Quite simply this is a visit to

Mecca, a pilgrimage to the most holy of holiest places. When I started this book in 2020 hajj was in its 1441 Hijri year. This number resonated with me because elsewhere in this book I explained that the total minutes in a day is 1440. So, if in 2020 it was 1441 in the Islamic calendar, this means that the year 2023 equals 1444 another interesting number. 1+4+4+4 = 13 the world's most unlucky number. The Islamic calendar is based on lunar cycles and the year is 11 days shorter than the Gregorian year and being an odd number, it changes from 10 to 11 days in difference on a cyclical basis and does so every 33 years. This in turn interestingly means the Hajj.

- The 13th Major Arcana card in the Tarot deck is one which represents death but is a symbolic end to something before a new beginning. While symbolically it also intimates reincarnation and the end of physical life, the same principle applies that if they are spiritually enlightened, they will move onto the next existence
- President Harry Truman ordered the atomic detonation, which was carried out 33 days later August 6, 1945, by the B-29 bomber Enola Gay on the Special Bombing Mission No. 13 over Hiroshima and three days later Nagasaki
- George W Bush is a cousin of the Queen Elizabeth, 13th cousin once removed whose family governed over the original 13 colonies of America
- 13 days after the Summer Solstice is Independence Day
- The American dollar has 13 stars, 13 strips on the dollar pyramid, 13 letters is Annuit Coeptis, 13 vertical bars on the shield, 13 leaves on the olive branch, 13 fruits and 13 numbers in Department of the Treasury seal
- Colombian Broadcasting Systems (CBS) 3, 2, 19 = 32-19 = 13
- Central Intelligence Agency (CIA) 3, 9, 1 = 3+9+1 = 13
- Department of Homeland Security (DHS) = 4, 8, 19, = 4x8-19 = 13
- 13 is a baker's dozen
- 13th sign of the Zodiac - Ophiuchus
- 13 recognised crosses – the *crux quadrata* - Greek cross, the *crux immissa* - Latin cross, the *crux commissa* - Tau cross, Tree of Life cross, 8 pointed cross – Maltese cross, Cross of St Aemilian of Cogolla, Coptic Cross, Trilobed cross, Russian Orthodox cross, Marian cross, Papal cross, San Damiano cross, St Andrew's cross

Do you see 13?

14

The fear of the number 14 is *Tetradecaphobia.*

- In a Biblical sense the number 14 comes up a lot, 14 plagues, 14 rams, 14 rains, 14 lambs and the most terrifying of all to me 14 wives
- Osiris was cut into 14 pieces and flung over Egypt

15

Fear of number 15 is *Quindecimphobia.*

- Hemingway said, *"I have drunk since I was 15, (alcohol not water) and few things have given me more pleasure"*
- Andy Warhol said, *"In the future, everyone will be world-famous for 15 minutes"*
- 15 is the magik squares numbers discussed elsewhere in this book
- José Ortega y Gasset (philosopher) said, *"A revolution only lasts 15 years, a period which coincides with the effectiveness of a generation"*. This is an interesting statement because a generation was always measured from when a female was born to when she first gave birth. In the last 20 years this has changed as females are waiting longer to give birth and a generation is now measured on average 25 years

16

Fear of 16 is *Hexadecaphobia.*

- Rosicrucian's believe that nature consists of 16 elements but first you must look at the base, which are:
    1. Fire The Salamanders
    2. Water The Undines
    3. Earth The Gnomes
    4. Air The Sylphs
- I also bring your attention to ancient Sanskrit as that BHUTI (Bhutus) were created from the elements and the beans that subsequently manifested were elementals, ghosts, goblins, imps, demons, phantoms, and elementaries. They were all nature spirits

17

Fear of the number 17 is *Heptadecaphobia.*

- 17 is the only prime number which is the sum of 4 consecutive primes
- On the 17th day of the second month the biblical flood began
- On the 17th day of the seventh month the water had diminished, and the ark rested on the mountains of Ararat
- The 17th day of the month the Greeks believed was the best day to cut wood for a boat
- There is no 17th floor in many buildings in Rome
- There is no seat 17 on Alitalia and Lufthansa flights. The reason is because 17 is XVII in Roman numerals, an anagram of VIXI, which in Latin translates as *'My life is over'*, implying that someone is dead
- Esteban Tuero, the Argentine driver who did a stint with the Minardi team in 1998, was notoriously superstitious and refused to run with cars carrying the numbers 13 or 17
- The bad luck No. 17 is also connected with the Roman XVII Legion which together with the XVIII and XIX annihilated in the Battle of the Teutoburg Forest 9 AD. Since then, the number 17 is considered unlucky in the Roman Army and were never again used to name military units

- The mystery schools consider this a highly spiritual number, the symbol of continuance
- 17th card of a Tarot deck is represented by the star and ties directly back to Osiris who raises from the dead on the 17th day of the month of ATHYR. In Egyptian mythology the number is for resurrection
- Exodus 20:1-17 in the Bible Commandments is explained in 17 verses and must be obeyed with respect towards God and his fellow man.
- 17 in the Islamic traditions is equally important. Initiates use 17 words in a daily prayer and 17 liturgical gestures. In the Shi'ite mysticism the number 17 was the amount of the dead brought back to life, also true for the ancient Greeks. We go back to the start of number 17, resurrection

18
The fear of 18 is *Octodecaphobia*, it is in the top 3 of the most feared numbers due to their calculation 6+6+6 = 18.

- The Hindu Mahabharata has 18 books
- The Jewish prayer Shemoneh Esreh refers to the original number of component blessings in the prayer (18)

19
The fear of the number 19 is *Enneadecaphobia.*

- 19 is an interesting number in relation to history which is a Satanic ritual day relating to fire, the fire God Moloch or Nimrod the Sun God also known as the Roman God of Saturn. This day was a major human sacrificial time demanding fire sacrifice with an emphasis particularly upon children. The 19th of April, the start of this 13th day ritual has direct relation to the 19th of April 2021 which is the date chosen by the British government to vaccinate every child up to 16 years of age
- Atomic Cycle Mitotic 19
- Tarot card 19 is The Sun, meaning happiness, success, and self-confidence
- 19 is a prime number

20
The fear of the number 20 is *Vigintiphobia.*

There is no occult relating to this number.

21
The fear of the number 21 is *Eikosihenaphobia.*

- Hoax (FR) = 21
- Fool (FR) = 21
- 21 in the mystery schools is several transformations which brings together the third cycle of seven or 3x7 = 21
- 21 in the Bible is associated with resurrection. It is still a tradition today that when a person reaches the age of 21, they reached the age of maturity and given the key to the door. This is an esoteric message that you are leaving your parents' home and

starting your own. In the Holy Bible there are 21 chapters in the Book of Judges, a considerable number as this is the start of your sinful nature for which God will hold you accountable

22

The fear of the number 22 *Vigintidyophobia.*

- Catch 22
- 22 Letters in the Hebrew alphabet
- 22 is the first human chromosome pair to be decoded
- Tarot card - the world card, voyage, flight, change of location, recompense, immigration, four beasts of the world. Interestingly Alistair Crowley called it the universe card under his rules of 777

23

The fear of number 23 *Eikositriophobia.*

- Blavatsky (FR) = 23
- Happy birthday - Some people may sing this to you, but in mathematics there is something called the 'happy birthday problem'. When I studied statistics, I found this area quite fascinating. Exploring the numbers regarding birthdate, it is called a bajillion probability. In a group of 23 the statistics is that there is a 50% chance of 2 people having the same birthday, in a group of 100 it is 99%. Therefore, with a group of 100 people, statistically we are related by numbers

24

- 24 time zones in the world. We cannot escape time, but the global time zones create occults.
  China covers 5 times zones geographically, but the Chinese government holds the entire country in 1 zone. If you walk across the border from China to Afghanistan, you must reset your watch by 3.30 hours. The Hawaiian Islands and Arizona (USA) do not follow Daylight Savings Time (DST). The state of Arizona has territories belonging to the indigenous early tribe of the Navajo Nation who follow DST. Inside the Navajo Nation you also have the Hopi Reservation which do not follow DST time, so you can see it getting messy. If you drive through Arizona from the boarder through Navajo and Hopi areas, you could change your clock 7 times

25

- Boris Johnson was not the first MP to ban Christmas *'oh no I hear you say'...* 25th December in ancient Rome was known as *"Dies Natalis Solis Invicti"* or the 'Birth of the Unconquered Sun'. If you are interested in history, then you would have heard of Oliver Cromwell who was a puritanical leader in the British parliament, he reigned as Lord Protector. During his reign, in 1647 Parliament passed an Ordinance that abolished Christmas day as a feast day and holiday
- Did you know Christmas 25th December was banned in America? In 1659 it was banned in Boston and those practicing would be fined 5 shillings. It was referred to a time of

sacrilege (stealer of sacred things – from sacred). In America Christmas did not become a federal holiday until 1870

- 25th December is said to be the date of the birth of Christ but wait let us look at mythology and other birthdays at this time in history:
    - Adonis Phoenicia 200 BC
    - Dionysus Greece 500 BC
    - Zarathustra Greece 1000 BC
    - Mithra Persia 600 BC
    - Nimrod
    - Tammuz Babylon 400BC
    - Osiris
    - Attis

26

The fear of number 26 *Eikosihexaphobia.*

There are no mystery school numbers related to 26.

27

- 3 x 3 x 3 = 27
- According to NASA, the Moon takes 27.322 days to orbit the Earth. Takes 27 days for the Moon to rotate once on its axis
- The 27 club is an age in which popular singers and artists die. The 27th prime number is 103 or 13. 13 and death go hand in hand

32

- America (FR) = 32
- 33 degreed of the Scottish Rite is the final illumination, the Sovereign Grand Inspector General
- Independence Day in America, the signed Declaration on the July 4th, 1776 holds a special place for American Freemasons. 7+4+1+7+7+6 = 32
- 32 is the God number in Freemasonry

33

- Gate (EO) 7+1+20+5 = 33
- Amen (EO) 1+13+5+14 = 33
- Seed (EO) 19+5+5+4 = 33
- RNA (EO) 18+14+1 = 33
- 33 levels of Freemasonry
- 33 sections in the United Nations flag
- 33° Mt Hermon (Nephilim)
- 33 Hz is called the Great Pyramid energy
- Arsenic Atomic number 33
- Masonry (FR) 4+1+1+6+5+9+7 = 33
- Bible (RFR) 7+9+7+6+4 = 33

- The Masonic Bible 20+8+5+13+1+19+15+14+9+3+2+9+2+12+5 = 137 the 33rd Prime number
- 33 vertebrae in the spine from the base of the body to the brain
- Jehovah (FR) 1+5+8+6+4+1+8 = 33
- Pineal (RFR) 2+9+4+4+8+6 = 33
- Gods Son (RFR) 2+3+5+8+8+3+4 = 33
- Lord God (RFR) 6+3+9+5+2+3+5 = 33
- James (RFR) as in King James 8+8+5+4+8 = 33
- Genesis (FR) 7+5+5+5+1+9+1 = 33
- The Word (RFR) 7+1+4+4+3+9+5 = 33 Genesis chapter 1 *'In the beginning was the Word, and the Word was God'*
- Sunday (RFR) 8+6+4+5+8+2 = 33
- Seventh (RFR) 8+4+5+4+4+7+1 = 33
- Seven Days (FR) 1+5+4+5+5+4+1+7+1 = 33
- Magic (FR) 13+1+7+9+3 = 33
- Hollywood (RFR) 1+3+6+6+2+4+3+3+5 = 33
- Left Eye (FR) 3+5+6+2+5+7+5 = 33
- Noah' Ark (FR) 5+6+1+8+1+1+9+2 = 33
- England (RFR) 4+4+2+6+8+4+5 = 33
- Brexit (FR) 2+9+5+6+9+2 = 33
- KKK (EO) 11+11+11 = 33
- LAPD (EO) 12+1+16+4 = 33
- NRA (EO) 14+18+1 = 33
- Akbar (EO) 1+ 11+2+1+18 = 33
- Serco Ltd (FR) 1+5+9+3+6+3+2+4 = 33
- Holocaust (FR) 8+6+3+6+3+1+3+1+2 = 33
- Person (FR) 7+5+9+1+6+5 = 33
- Police (FR) 7+6+3+9+3+5 = 33
- Bonesmen (FR) 2+6+5+5+1+4+5+5 = 33
- Society (FR) 1+6+3+9+5+2+7 = 33
- Secrecy (FR) 1+5+3+9+5+3+7 = 33
- False Flag (FR) 6+1+3+1+5+6+3+1+7 = 33
- Federal (FR) 6+5+4+5+9+1+3 = 33
- Order (FR) 6+9+4+5+9 = 33
- Priest (FR) 7+9+9+5+1+2 = 33
- Dow Jones (FR) 4+6+5+1+6+5+5+1 = 33
- People (FR) 7+5+6+7+3+5 = 33
- Believe (FR) 2+5+3+9+5+4+5 = 33
- Monsters (FR) 4+6+5+1+2+5+9+1 = 33
- Empire (RFR) 4+5+2+9+9+4 = 33
- Race War (FR) 9+1+3+5+5+1+9 = 33
- Mafia (RFR) 5+8+3+9+8 = 33
- Big Bang (FR) 2+9+7+2+1+5+7 = 33
- Name (EO) 14+1+13+5 = 33
- Desire (FR) 4+5+1+9+9+5 = 33

- New York (RFR) 4+4+4+2+3+9+7 = 33
- Jesus died aged 33 or spent 33 years on Earth
- 33° is the optimum temperature for the virus to mutate and destroy the human body from within and particularly with Pulmonary I L33 immune cells
- Bill Gates (FR) 2+9+3+3+7+1+2+5+1 = 33 he was involved with Microsoft for 33 years. Things get funky when you add up Melinda Gates = 47 Microsoft relocated to Redmond Washington also on the 47th parallel between the 2/18/86 and 2/26/86 which was the 47th day of the year. The original headquarters was in New Mexico the 47th state in the US and home to the Roswell incident of 1947
- London is divided into 32 boroughs, they are follows, Barking and Dagenham, Barnett, Bexley, Brent, Bromley, Camden, Croydon, Ealing, Enfield, Greenwich, Hackney, Hammersmith and Fulham, Haringey, Harrow, Havering, Hillingdon, Hounslow, Islington, Kensington and Chelsea, Kingston upon Thames, Lambeth, Lewisham, Merton, Newham, Redbridge, Richmond-upon-Thames, Southwark, Sutton, Tower Hamlets, Waltham Forest, Wandsworth, Westminster. In the City of London brings this to a total of 33
- Batman fans of the original series (1966) need to review the Series 1 33rd episode titled *'Fine Finny Fiends'* where it transpires that Bruce Wayne's father founded skull and bones. The 322 fraternity
- George Washington was a 33rd Degree Freemason within the Scottish Rite. The house of the temple the of Scottish Rite headquarters in Washington DC has 33 columns on the outside each being 33 feet tall. While it is in Oregon, numerically this is Americas 33rd state
- The original Microsoft logo had seven dots in each row, it also had 11 and 13 with some products but look closer at the logo itself can you now see 33?

- If you look at the date 4th July 1776 to 4th July 1945 equals 169 years, this time demonstrates the birth of America to the birth of the nuclear age. You will agree that it is a mere coincidence that it was carried out by the 33rd president. Fascinatingly that the 1st test coded named Trinity was carried out at the USAAF Alamogordo Bombing and Gunnery Range in White Sands New Mexico which is in the 33rd parallel. Both Nagasaki and Hiroshima run parallel to the 33rd parallel. And if you have ever heard of the Roswell incident you will know that this location in New Mexico is also on the 33rd parallel
- Pope John Paul I was in office for only 33 days prior to death
- German's famous Reichstag fire took place the 27th February 1933 or 2–27–33, 2+2+7 = 11 and 33
- In England there is a yearly celebration and procession known as the Lord Mayor Show, steeped in history and allegory with many things which are presented to the public in which they know nothing of. If ever you get the opportunity to watch this you will see

two stone benevolent, we call giants that represented Gog and Magog, also known as Gogmagog and Corenius. You would be forgiven for not knowing who these legendary representations are and why Henry V of England insisted they were in the parade. The story is that Diocletian the Roman Emperor was said to have 33 wicked daughters all of which he found husbands to control their unruly ways. The eldest daughter Alba organised her sisters into slitting the throats of all the husbands while they slept. Following this crime, they were all cast adrift in a single ship along with half a year's rations for survival and ended up on the Island of Albion (Great Britain). Here starts the correlation with the book of Enoch where giants lived upon the land and the legend has it that these females mated with them and created a population of half giants. Following the fall of Brute of Troy, the descendant of Aeneas fled to the same islands and killed Gogmagog with a single rock and sling and was rewarded Cornwall as his lands for his victory in this pre-Christianity legend. 33 has great significance to the City of London

- 33 turns are required to make a complete spiral sequence of human DNA
- Within the forgotten book of Enoch, it mentions Mount Hermon where the Nephilim came to Earth, which just happens to be 33° latitude and longitude. Babylon and Baghdad are both on or near the 33rd parallel, as are the pyramids
- The US went bankrupt in 1933 and Hitler rose to power as the German Chancellor in 1933

Some surprising connections and discoveries related to events found on the 33rd parallel:

  - The Bermuda Triangle - Flight 19
  - Phoenix Lights - 1993
  - Battle of Los Angeles - 1942
  - Roswell New Mexico - 1947
  - JFK Assassination - 1963
  - RFK Assassination - 1968
  - Death penalty allowed in each state & country along 33rd parallel
  - First atomic bomb test in Nevada
  - Atomic bomb dropped on Hiroshima & Nagasaki
  - The Great Pyramid
  - The Civil War began in Charleston, South Carolina in 1861

34

- D Day (EO) 4+4+1+25 = 34

36

- There are 36 oaths of loyalty within the Chinese Triad Societies

37

- There is a hidden pattern within the numbers:
  - 111, 1+1+1 = 3 and 3×37 = 111
  - 222, 2+2+2 = 6 and 6×37 = 222
  - 333, 3+3+3 = 9 and 9×37 = 333
  - 444, 4+4+4 = 12 and 12×37 = 444
  - 555, 5+5+5 = 15 and 15×37 = 555

  - 666, 6+6+6 = 18 and 18×37 = 666
  - 777, 7+7+7 = 21 and 21×37 = 777
  - 888, 8+8+8 = 24 and 24×37 = 888
  - 999, 9+9+9 = 27 and 27×37 = 999

38

- God Saturn (FR) 7+6+4+1+1+2+3+9+5 = 38
- Prayer (FR) 7+9+1+7+5+9 = 38
- Martial Law (FR) 4+1+9+2+9+1+3+3+1+5 = 38
- The King (FR) 2+8+5+2+9+5+7 = 38
- Pentagon (FR) 7+5+5+2+1+7+6+5 = 38
- Christmas (FR) 3+8+9+9+1+2+4+1+1 = 38
- English (FR) 5+5+7+3+9+1+8 = 38
- Killing (FR) 2+9+3+3+9+5+7 = 38

39

The mystery schools do not use this number.

40

- Biblically the number 40 is privation and trial in the Bible. Moses wandered in the desert for 40 years, there were 40 days of the great flood and Christ himself was tempted by the devil and spent 40 days in the wilderness
- -40° or 40° below is the only temperature that is the same in both Fahrenheit and Celsius
- There are 40 standard spaces on the monopoly board
- 40 hours in a standard working week
- A typical pregnancy 40 weeks
- Moses fasted twice for 40 days
- 40 days elapsed between resurrection and ascension of Jesus
- The Israelites stayed with the Philistine for 40 years

41

The mystery schools do not use this number.

42

- The angle of vision from the ground to see the full set of colours in a rainbow

43 - 44

The mystery schools do not have a use for this number.

45

The mystery schools do not have a use for this number. It is however related to geometry.

46

- Temple mount (FR) 2+5+4+7+3+5+4+6+3+5+2 = 46
- Sacrifice (FR) 1+1+3+9+9+6+9+3+5 = 46

- Children (FR) 3+8+9+3+4+9+5+5 = 46
- Southern Jurisdiction, a member who is a 32° Scottish Rite Mason for 46 months can be elected to Knight Commander of the Court of Honour and after 46 months as a KCCH then eligible for 33rd degree. The York Rite has 13 - 33+13 = 46

47

The mystery schools' number for confusion.

- Division (FR) 4+9+4+9+1+9+6+5 = 47
- Murdered (RFR) 5+6+9+5+4+9+4+5 = 47
- Agreement (RFR) 8+2+9+4+4+5+4+4+7 = 47
- Authority (FR) 1+3+2+8+6+9+9+2+7 = 47
- Cop (RO) 24+12+11 = 47
- Policeman (RFR) 2+3+6+9+6+4+5+8+4 = 47
- Republican (FR) 9+5+7+3+2+3+9+3+1+5 = 47
- Democrat (RFR) 5+4+5+3+6+9+8+7 = 47
- News (RO) 13+22+4+8 = 47
- Radio (EO) 18+1+4+9+15 = 47
- The CIA was created in 1947, government (RFR) 2+3+5+4+9+4+5+4+4+7 = 47

48 - 52

There are no mystery school numbers in relation to these.

53

- Rothschild (FR) 9+6+2+8+1+3+8+9+3+4 = 53
- Zuckerberg (FR) 8+3+3+2+5+9+2+5+9+7 = 53
- Joseph Biden (FR) 1+6+1+5+7+8+2+9+4+5+5 = 53
- Alzheimer's (FR) 1+3+8+8+5+9+4+5+9+1 = 53

54

- Number of squares on a Rubik's Cube
- Number of cards in a deck including the 2 Jokers

55

- It is the largest Fibonacci number to also be a triangular number
- A square pyramidal number (the sum of the squares of the integers one to 5) and a heptagonal number
- Mystery schools regard this as a life change and purpose number
- Rothschild (RFR) 9+3+7+1+8+6+1+9+6+5= 55
- Zuckerberg (RFR) 1+6+6+7+4+9+7+4+9+2 = 55
- Joseph Biden (RFR) 8+3+8+4+2+1+7+9+5+4+4 = 55
- Alzheimer's (RFR) 8+6+1+1+4+9+5+4+9+8 = 55

56

- Coronavirus (FR) 3+6+9+6+5+1+4+9+9+3+1 = 56
- Mind control (FR) 4+9+5+4+3+6+5+2+9+6+3 = 56

- Anthony Fauci (FR) 1+5+2+8+6+5+7+6+1+3+3+9 = 56
- Washington DC (FR) 5+1+1+8+9+5+7+2+6+5+4+3 = 56
- Unemployment (FR) 3+5+5+4+7+3+6+7+4+5+5+2 = 56
- 56 people signed the declaration of Independence

57

There is no use for this number in the mystery schools.

58

- Adding the first 7 prime numbers - 58
- Indigenous Native Americans believe this to be an unnatural number (in mathematics it is natural). Psalm 58, the prayer for God to punish the sinful
- Cricket - both batsmen must cross the 58 feet (18 meter) to score 1 run
- The number of counties in California
- 5+8 = 13 a number to be avoided in the West

59

There is no use for this number in the mystery schools.

60

- The 911 Pentagon building attack happened to be on the same day, 60 years after the ground broke to start construction
- 60 cubits is the height of Adam according to the Hadith from Sahih al-Bukhari

61

- Jeffrey Bezos (FR) 1+5+6+6+9+5+7+2+5+8+6+1 = 61
- Billionaire (FR) 2+9+3+3+9+6+5+1+9+9+5 = 61

62 - 65

There is no use for these numbers in the mystery schools.

66

- Trader (EO) 20+18+1+4+5+18 = 66
- Mankind (EO) 13+1+14+11+9+14+4 = 66
- Corona (EO) 3+15+18+15+14+1= 66
- 66 feet is the height of the Great Sphinx in Egypt, 240 feet in length
- 66 books in the Protestant Bibles
- 66° North and South of the Equator respectively is the Arctic and Antarctic circle's location
- 66th most abundant metal on Earth is mercury
- 666 is the squares of the first 7 prime numbers - $2^2+3^2+5^2+7^2+11^2+13^2+17^2 = 666$
- $T6.6 = T36 = 666$ is a number in triangulation or a repdigit
- $\sum_{n=1}^{666} 2n(-1)^n = 666$ interestingly this is the equation to work out the mathematics on a roulette wheel

67
There is no use for these numbers in the mystery schools.

68
World War I
- Began 28-7-1914 – 28+7+19+14 = 68

World War II
- Began 01-09-1939 – 1+9+19+39 = 68

Invasion of Ukraine
- Began 24-02-2022 – 24+2+20+22 = 68

    68
    68
    68

    So, you can see 666 but what of the 3 8's? 8+8+8 = 24, 2+4 = 6

69
The mystery schools have no use for this number.

70
- Janus worship (RFR) 8+8+4+6+8+4+3+9+8+1+9+2 = 70

71
- Catholic (RO) 3+1+20+8+15+12+9+3 = 71
- Babylon (RO) 2+1+2+25+12+15+14 = 71

72
- Bitcoin (EO) 2+9+20+3+15+9+14 = 72
- Money (EO) 13+15+14+5+25 = 72
- Dogecoin (EO) 4+15+7+5+3+15+9+14 = 72
- Solomon controlled 72 demons according to Red Freemasonry
- The Kabbalah '72 Names of God' formula, it is the formula Moses used to overcome the laws of nature
- 72 accomplices who helped Set kill his brother Osiris (according to some versions of the story)
- The Ars Goetia are the 72 elite demons of Hell (also known as the 72 Pillars)

73
The mystery schools have no use for this number.

74
- 74 this also relates to 7+4 = 11
- Simple (EO) 19+9+13+16+12+5 = 74
- English (EO) 5+14+7+12+9+19+8 = 74
- Gematria (EO) 7+5+13+1+20+18+9+1 = 74

- Jesus (EO) 10+5+19+21+19 = 74
- Messiah (EO) 13+5+19+19+9+1+8 = 74
- Gospel (EO) 7+15+19+16+5+12 = 74
- Cross (EO) 3+18+15+19+19 = 74
- Lucifer (EO) 12+21+3+9+6+5+18 = 74
- Drown (EO) 4+18+15+23+14 = 74
- Occult (EO) 15+3+3+21+12+20 = 74
- Muhammad (EO) 13+21+8+1+13+13+1+4 = 74
- Masonic (EO) 13+1+19+15+14+9+3 = 74
- London (EO) 12+15+14+4+15+14 = 74
- Tarot (EO) 20+1+18+15+20 = 74
- Weapon (EO) 23+5+1+16+15+14 = 74
- Nuclear (EO) 14+21+3+12+5+1+18 = 74
- Jewish (EO) 10+5+23+9+19+8 = 74
- Heavens (EO) 8+5+1+22+5+14+19 = 74
- The King (EO) 20+8+5+11+9+14+7 = 74
- Ruler (EO) 18+21+12+5+18 = 74
- Theism(EO) 20+8+5+9+19+13 = 74
- Joshua (EO) 10+15+19+8+21+1 = 74
- Son God (EO) 19+15+14+7+15+4 = 74
- Parables (EO) 16+1+18+1+2+12+5+19 = 74
- God and a Man (EO) 7+15+4+1+14+4+1+13+1+14 = 74
- Evil God (EO) 5+22+9+12+7+15+4 = 74
- The Key (EO) 20+8+5+11+5+25 = 74
- Connect (EO) 3+15+14+14+5+3+20 = 74
- Rapper (EO) 18+1+16+16+5+18 = 74
- Jury (EO) 10+21+18+25 = 74
- Energy (EO) 5+14+5+18+7+25 = 74
- Preacher (EO) 16+18+5+1+3+8+5+18 = 74
- Demon is mentioned 74 times in the Bible
- Trauma (EO) 20+18+1+21+13+1 = 74
- Killing (EO) 11+9+12+12+9+14+7 = 74
- Synchronicity (FR) 1+7+5+3+8+9+6+5+9+3+9+2+7 = 74
- Now the new normal (FR) 5+6+5+2+8+5+5+5+5+6+9+4+1+3 = 74
- Fiendish (EO) 6+9+5+14+4+9+19+8 = 74
- Horns (EO) 8+15+18+14+19 = 74
- Old goat (EO) 15+12+4+7+15+1+20 = 74
- Tempt (EO) 20+5+13+16+20 = 74
- Yellow brick road (FR) 7+5+3+3+6+5+2+9+9+3+2+9+6+1+4 = 74
- The Global Vaccine Summit was hosted in London in 2020 and exceeded the target of $7.4 billion

75 - 76

The mystery schools have no use for this number.

77

- Police department (FR) 7+6+3+9+3+5+4+5+7+1+9+2+4+5+5+2 = 77
- Police officer (FR) 7+6+3+9+3+5+6+6+6+9+3+5+9 = 77
- False flag shooting (FR) 6+1+3+1+5+6+3+1+7+1+8+6+6+2+9+5+7 = 77
- American Airlines flight 77 had an impact with the Pentagon, which is 77 feet tall on the 77th meridian west exactly 77 minutes after taking off at 8:20 and crashed at 9.37 am
- World Trade Centre (FR) 5+6+9+3+4+2+9+1+4+5+3+5+5+2+9+5 = 77
- September Eleventh (FR) 1+5+7+2+5+4+2+5+9+5+3+5+4+5+5+2+8 = 77
- Power (EO) 16+15+23+5+18 = 77
- Wake up (EO) 23+1+11+5+21+16 = 77
- United States (RFR) 6+4+9+7+4+5+8+7+8+7+4+8 =77

78 - 80

The mystery schools have no use for these numbers, but it has been 80 years between a Covid passport and the Nazis yellow star.

81

- Mask up (EO) 13+1+19+11+21+16 = 81
- Masons (EO) 13+1+19+15+14+19 = 81
- Ritual (EO) 18+9+20+21+1+12 = 81

82 - 85

Mystery schools have no use for these numbers.

86

- Hallelujah means 'praise the Lord' in Hebrew it adds up to 86 but strangely enough in the Greek it adds up to...you guessed it 86.

87 - 92

The mystery schools have no use for these numbers.

93

- Saturn (EO) 19+1+20+21+18+14 = 93

94

- Coronavirus Pandemic (FR) 3+6+9+6+5+1+4+9+9+3+1+7+1+5+4+5+4+9+3 = 94
- Alien (RO) 26+15+18+22+13 = 94
- Starchild (EO) 19+20+1+18+3+8+9+12+4 = 94
- Project Blue Beam leaked to the public in 1994
- Ratio of 9:4 is the proportion that is most visible within the Parthenon, the ratio of 3:2, the musical fifth, is key for its proportioning. The ratios of length, height, width of the outer temple to the length, height, width of the inner temple (cella) is all in the ratio of 3:2. It is 101 feet wide.

99

- Firmament (EO) 6+9+18+13+1+13+5+14+20 = 99

- Creators (EO) 3+18+5+1+20+15+18+19 = 99
- Ascension (EO) 1+19+3+5+14+19+9+15+14 = 99
- Another interesting fact is that Prince Philip died in April on the 4th month of 2021. Look at the numbers you cannot see. He was 99 years of age, he died at 9 am on the 99th day of the year on the 9th of April. This gives you 9+9+9+9+9+9 = 54, 5+4 = 9 and looking at it another way 6 6s is 666666 balance the number 666-666 = 0 either way that is a lot of 9s and a hell of a yet another group of coincidence
- 99 names of Allah, Sahih Al-Bukhari - Book 50 Hadith 894 "*Allah has ninety-nine names i.e. one-hundred minus one, and whoever knows them will go to Paradise*".

100

The mystery schools have no use for this number.

101

- Gary Patrick Fraughen (FR) 7+1+9+7+7+1+2+9+9+3+2+6+9+1+3+7+8+5+5 = 101
- Christian (EO) 3+8+18+9+19+20+9+1+14 = 101
- Gods will (EO) 7+15+4+19+23+9+12+12 = 101
- Tree of Life (EO) 20+18+5+5+15+6+12+9+6+5 = 101
- Allah (RO) 26+15+15+26+19 + 101
- Google (RO) 20+12+12+20+15+22 = 101
- Walt Disney World Resort (FR) 5+1+3+2+4+9+1+5+5+7+5+6+9+3+4+9+5+1+6+9+2= 101
- Red Cross (EO) 18+5+4+3+18+15+19+19 = 101
- 101 - the telephone number to call the police for non-emergency issues in the UK
- Church (RO) 24+19+6+9+24+19 = 101
- 101 - Neo's apartment number in the Matrix film
- Gematria Code (EO) 7+5+13+1+20+18+9+1+3+15+4+5 = 101

102

- Wuhan China (EO) 23+21+8+1+14+3+8+9+14+1 = 102
- Social distancing (RFR) 8+3+6+9+8+6+5+9+8+7+8+4+6+9+4+2 = 102
- Babylon mother of harlots (FR) = 102
- Jesus Christ will be defeated (FR) = 102
- The fallen God of the last days (FR) = 102
- Proven to be a psychopath (FR) = 102
- An insufferable narcissist (FR) = 102
- Bloodlines rule the planet (FR) = 102

103 - 106

A mystery schools have no use for these numbers.

107

- Military (EO) 13+9+12+9+20+1+18+25 = 107
- Satan's name (EO) 19+1+20+1+14+19+14+1+13+5 = 107
- Devil Satan (EO) 4+5+22+9+12+19+1+20+1+14 = 107
- God on Earth (EO) 7+15+4+15+14+5+1+18+20+8 = 107

108

- The average distance between Earth to the Sun is approx. 108 x the diameter of the Sun
- The average distance between Moon and the Earth is approx. 108 x the Moon's diameter

109

- The Sun is 109 x the diameter of the Earth across

110

- The Twin Towers (part of the World Trade Center Complex) had 110 floors each
- Martial Law (EO) 13+1+18+20+9+1+12+12+1+23 = 1 10
- Centre for disease control (FR) = 110, this relates to number 101 and the *blood and zombies* chapter in this book

111

- New York (EO) 14+5+23+25+15+18+11 = 111
- Vatican Hill (EO) 22+1+20+9+3+1+14+8+9+12+12 = 111
- Mecca Saudi Arabia (EO) 13+5+3+3+1+19+1+21+4+9+1+18+1+2+9+1 = 111
- La Sainte Bible (EO) 12+1+19+1+9+14+20+5+2+9+2+12+5 = 111
- Accept hand chip (EO) 1+3+3+5+16+20+8+1+14+4+3+8+9+16 = 111
- RFID Scanner (EO) 18+6+9+4+19+3+1+14+14+5+18 = 111
- Satan's RFID (EO) 19+1+20+1+14+19+18+6+9+4 = 111
- Vaccination (EO) 22+1+3+3+9+14+1+20+9+15+14 = 111
- Computer (EO) 3+15+13+16+21+20+5+18 = 111
- Necromancy (EO) 14+5+3+18+15+13+1+14+3+25 = 111
- Witchcraft (EO) 23+9+20+3+8+3+18+1+6+20 = 111
- Mark of beast (EO) 13+1+18+11+15+6+2+5+1+19+20 = 111
- Illusion(EO) 9+12+12+21+19+9+15+14 = 111
- Insanity (EO) 9+14+19+1+14+9+20+25 = 111
- Slaughter (EO) 19+12+1+21+7+8+20+5+18 = 111
- A covid vaccine (EO) 1+3+15+22+9+4+22+1+3+3+9+14+5 = 111
- Image of Satan (EO) 9+13+1+7+5+15+6+19+1+20+1+14 = 111
- Bio implant (EO) 2+9+15+9+13+16+12+1+14+20 = 111
- A satanic mark (EO) 1+19+1+20+1+14+9+3+13+1+18+11 = 111
- Beast test (EO) 2+5+1+19+20+20+5+19+20 = 111
- Receive a mark (EO) 18+5+3+5+9+22+5+1+13+1+18+11 = 111
- The hand or head (EO) 20+8+5+8+1+14+4+15+18+8+5+1+4 = 111
- Forehead sign (EO) 6+15+18+5+8+5+1+4+19+9+7+14 = 111
- E-identity (EO) 5+9+4+5+14+20+9+20+25 = 111
- People Sin (EO) 16+5+15+16+12+5+19+9+14 = 111
- Son of Sin (EO) 19+15+14+15+6+19+9+14 = 111
- Hardened Heart (EO) 8+1+18+4+5+14+5+4+8+5+1+18+20 = 111
- Wicked will (EO) 23+9+3+3+11+5+4+23+9+12+12 = 111
- Book of the Dead (EO) 2+15+15+11+15+6+20+8+5+4+5+4+5+1+4 = 111

- An absence of God (EO) 1+14+1+2+19+5+14+3+5+15+6+7+15+4 = 111
- A third wave (EO) 1+20+8+9+18+4+23+1+22+5 = 111
- Beast is here (EO) 2+5+1+19+20+9+19+8+5+18+5 = 111
- Ancient gods (EO) 1+14+3+9+5+14+20+7+15+4+19 = 111
- Heaven and hell (EO) 8+5+1+22+5+14+1+14+4+8+5+12+12 = 111
- There is a god (EO) 20+8+5+18+5+9+19+1+7+15+4 = 111
- Humanity (EO) 8+21+13+1+14+9+20+25 = 111
- Secrecy (RO) 8+22+24+9+22+24+2 = 111
- Dollar Crash (EO) 4+15+12+12+1+18+3+18+1+19+8 = 111
- Sheep Asleep (EO) 19+8+5+5+16+1+19+12+5+5+16 = 111
- Fluoridate (EO) 6+12+21+15+18+9+4+1+20+5 = 111
- Melinda A Gates (EO) 13+5+12+9+14+4+1+1+7+1+20+5+19 = 111
- Internet (RO) 18+13+7+22+9+13+22+7 = 111
- Mandatory (EO) 13+1+14+4+1+20+15+18+25 = 111
- Forced (RO) 21+12+9+24+22+23 = 111
- Satanic mask (EO) 19+1+20+1+14+9+3+13+1+19+11 = 111
- Evil masks (EO) 5+22+9+12+13+1+19+11+19 = 111
- The face masks (EO) 20+8+5+6+1+3+5+13+1+19+11+19 = 111
- Energy 3, 6 and 9 (EO) 5+14+5+18+7+25+3+6+1+14+4+9 = 111
- Dark matter (EO) 4+1+18+11+13+1+20+20+5+18 = 111
- 3,6 and 9 occult (EO) 3+6+1+14+4+9+15+3+3+21+12+20 = 111
- 3,6,9 harvest (EO) 3+6+9+8+1+18+22+5+19+20 = 111
- Energy and 369 (EO) 5+14+5+18+7+25+1+14+4+3+6+9 = 111
- Electric field (EO) 5+12+5+3+20+18+9+3+6+9+5+12+4 = 111
- Vaccination (EO) = 111 but (ES) 132+6+18+18+54+84+6+120+54+90+84 = 666
- Pizza gate (EO) = 111 but in (ES) 96+54+156+156+6+42+6+120+30 = 666
- Corrupt (EO) = 111 but (ES) 18+90+108+108+126+96+120 = 666
- Santa Claus (EO) 19+1+14+20+1+3+12+1+21+19 = 111
- World War I ended on the 11th month of the 11th day on the 11th hour
- World War I was started by all common accounts with the assassination of the Archduke Franz Ferdinand in a vehicle whose registration number was A III 118
- Child snuff film collectors (FR) = 111
- Internal revenue service (FR) = 111
- illuminati Corporation (FR) = 111
- Millions of dollars income (FR) = 111
- The serpent and the rainbow (FR) = 111
- Slave ship (EO) = 111 but (ES) 114+72+6+132+30+114+48+54+96 = 666

113

- Illusions (RO) = 113
- Politics (RO) = 113
- Not honest (RO) = 113
- Fiction (RO) = 113
- Bullshit (RO) = 113
- Divorce (RO) = 113

- Mainstream (EO) = 113
- Broadcasting (EO) = 113
- Not true (EO) = 113
- Dishonest (EO) = 113
- Not factual (EO) = 113

118

- Sacrificial Lamb (EO) 19+1+3+18+9+6+9+3+9+1+12+12+1+13+2 = 118
- Sacrifice the children (RFR) = 118

Bible numbers:

- Psalm 118 is the middle chapter of the entire Bible
- Psalm 117, the shortest chapter is before Psalm 118
- Psalm 119, is the longest chapter after Psalm 118
- The Bible has 594 chapters before and after Psalm 118
    - If you add up all the chapters except Psalm 118 you get a total of 1188. 1188 or Psalm 118 verse 8 is the middle verse of the entire Bible. Interestingly Psalm 118:8 reads*"It is better to take refuge in the LORD than to trust in man"*.

121

- Coronavirus outbreak (RFR) = 121
- Bill and Melinda Gates foundation (FR) = 121
- Revelation (EO) = 121
- Blood sacrifice(EO) = 121

127

- The Oval Office (EO) = 127
- Vatican City (EO) = 127
- Buckingham Palace (EO) = 127
- Symbolism (EO) = 127

133

- Flat earthers (EO) = 133

137

- Morals and Dogma (EO) 13+15+18+1+12+19+1+14+4+4+15+7+13+1 = 137
- Eye of Horus (EO) 5+25+5+15+6+8+15+18+21+19 = 137
- In the Beginning (EO) 9+14+20+8+5+2+5+7+9+14+14+9+14+7 = 137
- Church of Satan (EO) 3+8+21+18+3+8+15+6+19+1+20+1+14 = 137
- Finding Jesus (EO) 6+9+14+4+9+14+7+10+5+19+21+19 = 137
- Authority (EO) 1+21+20+8+15+18+9+20+25 = 137
- Mind Control (EO) 13+9+14+4+3+15+14+20+18+15+12 = 137
- Bill of Rights (EO) 2+9+12+12+15+6+18+9+7+8+20+19 = 137
- Scottish Rite Freemasonry (RFR) 8+6+3+7+7+9+8+1+9+9+7+4+3+9+4+4+5+8+8+3+4+9+2 = 137

- Royal Family (EO) 18+15+25+1+12+6+1+13+9+12+25 = 137
- Royal Wedding (EO) 18+15+25+1+12+23+5+4+4+9+14+7 = 137
- Westminster paedophile ring (RFR) 4+4+8+7+5+9+4+8+7+4+9+2+8+4+5+3+2+1+9+6+4+9+9+4+2 = 137
- Brexit is the new world order (RFR) 7+9+4+3+9+7+9+8+7+1+4+4+4+4+4+3+9+6+5+3+9+5+4+9 = 137
- Celebrity Death (EO) 3+5+12+5+2+18+9+20+25+4+5+1+20+8 = 137

139

- Biblically Psalm 139 is regarding DNA

166

- September eleven (EO) 19+5+16+20+5+13+2+5+18+5+12+5+22+5+14 = 166
- The White House (EO) 20+8+5+23+8+9+20+5+8+15+21+19+5 = 166
- One World Order (EO) 15+14+5+23+15+18+12+4+15+18+4+5+18 = 166
- Annuit Coeptis (EO) 1+14+14+21+9+20+3+15+5+16+20+9+19 = 166
- The nine eleventh (EO) 20+8+5+14+9+14+5+5+12+5+22+5+14+20+8 = 166
- Bin Laden nine eleven (EO) 2+9+14+12+1+4+5+14+14+9+14+5+5+12+5+22+5+14 = 166
- Revealing number (EO) 18+5+22+5+1+12+9+14+7+14+21+13+2+5+18 = 166

222

- World Economic Forum (EO) = 222
- Wuhan Coronavirus (EO) 23+21+8+1+14+3+15+18+15+14+1+22+9+18+21+19 = 222
- Order Out of Chaos (RO) = 222
- As above so below (RO) = 222
- New York New York (EO) 14+5+23+25+15+18+11+14+5+23+25+15+18+11 = 222
- Mandatory (EO) 13+1+14+4+1+20+15+18+25 = 111
- Vaccination (EO) 22+1+3+3+9+14+1+20+9+15+14 = 111

  add the 2 above names together and you get:
  13+1+14+4+1+20+15+18+25 and 22+1+3+3+9+14+1+20+9+15+14 = 222

303

- For the festival of saturnalia (EO) = 303
- Never suffer from regret (RO) = 303
- The serpents are lawless (RO) = 303

311

- 311 and 911 are both numbers to call authorities. Where 911 is for an emergency, 311 is for non-emergencies

322

- SpaceX launched its 1st private rocket to dock with the International Space Station on 30th May 2020 at 3.22 PM Eastern Time
- Adam Weishaupt died on 18th November 1830 which interestingly is the 322nd day of the year

- The Skull and Bones, known as Order 322, the number also appears on the society's insignia and on the building at Yale University
- A Paper Plane and Jenga Blocks = 223 or reversed 322
- New World Order the new normal (EO) = 322
- Thank you for the warning lord (EO) = 322
- The end of the world as we know it (EO) = 322
- I briefly mentioned 322 in the "stopper 1". Genesis 3:22 states *"and the Lord God said the man has now become like one of us, knowing good and evil"*. Take this statement and turn it into numeracy (RFR) = 322
- The 202nd minute of a day is 3-22 (3 hours and 22 minutes)
- The Georgia Guidestones erected on March 22 (322) in 1980

333

Fear of number 333 is Trikosioitriakontatrophobia.

- Corona is in the vaccine (RO) = 333
- Corona vaccine deaths (RO) = 333
- Moderna covid vaccine (RO) = 333
- Pope Francis is the 266th Bishop in Rome. Nothing odd in that number but let us put it into English Ordinal (EO) code 'two hundred sixty six pope' 20+23+15+8+21+14+4+18+5+4+19+9+24+20+25+19+9+24+16+15+16+5 = 333
- The George Washington Masonic National Memorial building in Virginia stand 333 feet tall
- PCR Test (RS) = 333

Mystery schools and 3333:

- We see the heavy use of 333 as a harm number, so why did the mystery schools use it? This falls back to scripture and the crucifying of Jesus which is filled with the number 3333. Jesus died at the age of 33, in 33 AD on the 3rd of April

365

- Days in the year
- How many times "do not be afraid "is mentioned in the Bible

444

- Jesus (ES) 60+30+114+126+114 = 444
- Lucifer (ES) 72+126+18+54+36+30+108 = 444

666

Hexakosioihexekontahexaphobia: an individual who fears the number 666.

- The sum of the squares of the first 7 primes is 666:
  - $666 = 2^2 + 3^2 + 5^2 + 7^2 + 11^2 + 13^2 + 17^2$
- The triplet (216, 630, 666) is a Pythagorean triplet. This fact can be rewritten in the following nice form:
  - $(6x6x6)^2 + (666 - 6x6)^2 = 666^2$

- A well-known remarkably good approximation to pi is 355/113 = 3.1415929...If one part of this fraction is reversed and added to the other part, we get:
  - 553 + 113 = 666
- From Martin Gardner's "*Dr. Matrix*" columns, Dewey Decimal Classification number for 'Numerology' is 133.335. If you reverse this and add them together, you get 133.335 + 533.331 = 666.666
- SevenDust was a virus that infected the Macintosh computer in 1998 and it sometimes goes by the name 666
- In spirituality it is associated with balance and in China considered to be incredibly lucky and is often on shop signs
- 66,616 mph is the Earth's average orbital speed
- 66.6° is the remaining measurement of the Earth's axial tilt at 23.4°
- The thirty-sixth triangular number, declared the number of the beast = 666
- A numeral thrice repeated, the triangular expansion of thirty-six = 666
- Revelation of Jesus to John, 13th chapter, 8th verse = 666
- Wisdom: *"Let him that hath understanding count the number of the beast"* = 666
- The Gospels, the Acts of the Apostles, the Epistles, and the Apocalypse = 666
- The Yorkshire ripper (EO) = 243 invert it +342 = 666
- 1 Kings 10:14 *"Now the weight of gold that came to Solomon in one year was six hundred threescore and six talents of gold"*. 666 mentioned
- Apple priced their first computer at $666.66 in 1976
- Bill Gates old company Microsoft filled a patent on 20th June 2019 which was published on the 26th March 2020, number WO2020060606 to monitor future cryptocurrency system using body activity
- Pope Francis birthday was in 2001 was the 17th December, the Coronavirus pandemic was declared on the 11th March 2020 which totalled 6660 days
- Saturn's hexagonally north pole is six sided, it is a six-pointed polygon representing the cube containing six triangles (a hex) and the 6th planet. Interestingly its orbital distance is 1426, 666, 422 km
- The Earth's axis in its orbital inclination around the Sun is 66.6°
- Image of Satan (ES) = 666
- Necromancy (ES) = 666
- Witchcraft (ES) 138+54+120+18+48+18+108+6+36+120 = 666
- build back better (EO) 2+29+9+12+4+2+1+3+11+2+5+20+20+5+18 = 135, carry out a mirror calculation and you get 135+531 = 666. B in (EO) = 2, there are 3 Bs which add to 6. 6uild 6ack 6etter has a pattern

<u>National - Aeronautics - and - Space - Administration</u> (NASA):

- National (RO) 13+26+7+18+12+13+26+15 = 130
- Aeronautics (RO) 26+22+9+12+13+26+6+7+18+24+8 = 171
- And (RO) 26+13+23 = 62
- Space (RO) 8+11+26+24+22 = 91
- Administration (RO) 26+23+14+18+13+18+8+7+9+26+7+18+12+13 = 212
  - Total for NASA taking on the above totals is 130+171+62+91+212 = 666

As mentioned in a previous chapter Rome used to use 6 alphabet letters D=500 C=100 L=50 X=10 V=5 I=1:

D=500
C=100
L= 50
X= 10
V= 5
I= 1 +

666

Interestingly, if we look at these calculations further there is a 3 pattern. Line them up and here they are. (On the 1776 American Dollar bill is the calculation as above so below MDCCLXXVI).

M	C	X
D C	L X	V I
--------------------	--------------------------	-------------------------
D C	L X	V I
M	C	X
D + C = 600	L + X = 60	V + I = 6 hidden 666

The American Standard Code for Information Interchange or ASCII is the numerical representation of a character. Let us look at that code:

- Bill Gates 66+73+76+76+71+65+84+69+83+3 = 666
- MS Dos 6.21 77+83+45+68+79+83+32+54+46+50+49 = 666
- Windows 95 87+73+78+68+79+87+83+57+53+1 = 666

There are bar codes placed on all commercially viable products bought and sold, there is even one placed on this book and if you purchased from Amazon there are 2 for there is a sticker over the original ISBN. Amazon dictate that the original ISBN numbers be covered so they go out from the Amazon Seller Central without the code registered with the British library.

There are 3 categories on the barcode, 1 the universal credit code, 2 the company code and 3 the end code. Now look carefully at the barcode again there is a double line to start, a double line to finish, and a double line in the middle on the vertical striking themselves. These double lines represent 6, and as there are three of them, they denote 666. You are now asking why is this important? I draw your attention to Revelation 13:17 *"And that no man might buy or sell, save he that had the mark or the name of the beast or the number of his name"*. Is this a direct warning that at the end of days biblically speaking all nations will be trading under the number 666. I think you can see the correlation whether you are religious or not, it is worth pondering. Another unknown fact is that these databank codes were originally stored at the

United States Embassy building in Brussels and the name of the data storage system another coincidence it was called "the beast".

Numbers 1 to 36 calculate to 666 in the mystery schools, so 36 is the veiled reference to 666.

The mezuzah is an interesting worship within the Jewish faith and one of the most visible symbols of Judaism. Constructed within an outer casing, with a piece of parchment with specific Hebrew verses from the Torah or the Five Books of Mosses. Placed on the right of a door frame as you enter a property it is 666% of its height from the floor. Sephardi Jews place them vertically while Ashkenazim Jews place them at an angle of 30% or less from the vertical. They are opened and checked every 7 years for perfect condition.

777

- On the 7th July 2005, the London Tube bombings took place. 7/7/2+5 = 7 = 777
- Aleister Crowley's book *"777 and Other Qabalistic Writings"* was the addition he revealed to us *"do what thow wilt"* a number created purposely by misspelling Babalon, therefore 777
- In the esoteric schools this is a number for revenge and death. Lamech in the bible said onto his wives Adah and Zillah *"If Cain shall be avenged sevenfold, truly Lamech seventy and sevenfold"*. If revenge is 77, it is 11 times Cains 7
- The assassination of JFK happened in Dallas along Highway 77 and Masonic Grand Lodge of Oklahoma in Guthrie is along US route 77

Timothy McVeigh rented a Ryder truck along 77 and was also taken to jail along Highway 77 after his arrest. Both the Waco massacre and the Oklahoma City bombing were on Highway 77. Not only did they both occur on the same day 19th April 1993 for Waco and 1995 for Oklahoma but also the 2004 Chicago Sears Tower bombing plot and was exactly 444 days after the sears tower on 7/7/2005 (2+5 = 7) or 777. Are you not convinced? The London underground bombings were 77 with 52 victims being the outcome 5+2 = 7. Now we go on to the date 11th July 2006 in Mumbai where a terrorist train bomb killed 209 commuters or (2+9 = 11) using 7 bombs spaced exactly 11 minutes apart.

911

- When Jimmy Savile was buried on the 9th November 2011, his coffin was tilted to 45° angle. 4+5 = 9. 9th November = 9-11

1776

- The angular diameter of the Moon is 0.5° ... 888/0.5 = 1776
- The flood took place 1656 years from Adam... There have been 120 jubilees (50-year cycles) from Adam to 2000 AD. 1656 + 120 = 1776
- There are 1728 cubic inches in one cubic foot. Ersatz Israel became a nation again in 48...1728 + 48 = 1776
- 1776 ft, the exact height of the One World Trade Centre that replaced the Twin Towers

Chapter 5 – Coronavirus, Shakespeare and Faces

This chapter was a difficult one to navigate and descends into many rabbit holes, it is best read in one sitting for you to appreciate the threads I leave for you to explore. While we live in a mathematical universe, the secret societies use these numbers and they have infiltrated the belief systems to a level far beyond most people's appreciation. Words have direct correlation to numbers, let me show you how words can be backed into numbers and so I reveal a whole new world for you to discover.

Over the past two years the word Coronavirus has become the most frequently used set of letters typed into the Internet. Governments love fear, for its use keeps people in control with a deep-seated low resonance and vibration. Here is the year-by-year number list of Government fear narratives used against the world population while I was growing up:

- 1960s No more oil in 10 years
- 1970s Another ice age in 10 years
- 1990s The ozone layer will be destroyed in 10 years
- 1999 Ice flows will be gone in 10 years
- 2001 Y2K will destroy everything and kill us all
- 2001 Anthrax will kill us all
- 2002 W. Nile virus will kill us all
- 2003 SARS will kill us all
- 2005 Bird flu will kill us all
- 2006 E. coli will kill us all
- 2008 Financial crash will kill us all
- 2009 Swine flu will kill us all
- 2012 The MAYAN calendar will end and kill us all
- 2013 North Korea will kill us all
- 2014 Ebola will kill us all
- 2015 Isis will kill us all
- 2016 Zika virus will kill us all
- 2018 Global warming will kill us all then climate change will kill us all
- 2019 CO2 will kill us all
- 2020 Coronavirus will kill us all
- 2021 The vaccine will SAVE us all

There was a change in the narrative last year, it is not fear anymore, the government script has changed to HOPE. The question is for what end? Look at the word Terrorism, noun / TER – ror – ISM definition, 'the systematic use of terror especially as a means of coercion'.

5 countries refused to roll out the vaccine with no mandated lockdown and these were:

1. Haiti
2. Ivory Coast
3. Burundi
4. Tanzania
5. Swaziland

The leaders of all these countries are now dead, with most in suspicious circumstances.

- Jovenel Moise - Haiti - Died 07th July 2021 (Assassinated)
- Hamed Bakayoko - Ivory Coast - Died 10th March 2021
- Pierre Nkurunziza – Burundi – Died 8th June 2020
- John Magufuli – Tanzania – Died 17th March 2021
- Ambrose Mandvulo Dlamini – Swaziland – Died 13th December 2020

Following the deaths of these leaders they have all been replaced by members who trained with the *World Economic Forum* and the vaccine rolls outs in these places has now been embarked upon.

Population Numbers

In 2020 the USA who hold the biggest (not oldest) birth and death records on Earth announced 1 birth every 8 seconds but inversely 1 death every 12 seconds. They had a net gain of one soul every 26 seconds. For recreationists this is an issue for where are all these extra souls coming from, as there would be a mathematical imbalance with reincarnation.

The recording of humanity is complicated, but basic global calculations are as follows (including future projections):

Year	**Years elapsed**	**Population/Billion**
1804	data unavailable	1
1929	123	2
1960	33	3
1974	14	4
1987	13	5
1999	12	6
2011	12	7
2023	12	8
2037	14	9
2057	20	10

The world's population numbers are increasing or, so we are told, and the mantra is that it is unsustainable according to the renowned science specialist Greta the DOOM GOBLIN.

If we consider over population but from the standpoint that there are circa 7.3 billion humans on the planet, and we gave each of them a 2000 $ft.^2$ house in Russia which could accommodate all of them. The country is 6.6 $mi.^2$ would hold a total of 9.2 billion houses, so population numbers no longer seem an issue.

In the British newspaper the Telegraph, dated 25th October 2007 Boris Johnson (present UK Prime Minister) said the following:

"The world's population is now 6 .7 billion, roughly double what it was when I was born. If I live to be in my mid-eighties, then it will have trebled in my lifetime. The UN last year revised its forecasts upwards, predicting that there will be 9.2 billion people by 2050, and I simply

cannot understand why no one discusses this impending calamity, and why no world statesmen have the guts to treat the issue with the seriousness it deserves". I find this statement interesting for nobody knows how many children he has beyond the 6 that have been confirmed.

In 2020 an estimated 7.1 to 7.8 billion (some un-registered) people walked the Earth. But wait, something else happened which means I must bring you back to a monument mentioned elsewhere in this book and that is the Georgia Guidestones and its first inscribed commandment. *"Maintain humanity under 500 million in the perpetual balance with nature"*.

If in 2020 we remove 6 billion, 6 million and 6 hundred thousand (666) from the population, we are left with circa 500 million. On the 22/3/2022 at 2.22-20 or 222=6 222=6 222=6 or 666. The monument turned 40 years of age. Take 666 ÷ 40 = 16.65 and 5+1 = 6 so the 666 appears again.

2020 appears to be a year chosen due to the "elites" first commandment which triggered all the subsequent chaos. As of November 2021, data suggests that over half the world's population has now been vaccinated but as a matter of record in the years to come read this book knowing I as the author believe something far more sinister and malevolent is happening.

Does CO2=PxSxExC or P x S x E x C = CO2 mean anything in mathematical terms to you? Well, it should. It relates to carbon dioxide output. In the equation, P = population; S = services used by the people; E = the energy needed to power those services; and C = the carbon dioxide created by that energy. This is a population equation pushed by a Mr Bill Gates. He points out that scientists are calling for an 80 % drop in carbon emissions by 2050 (and a total end by 2100) to stave off the most dramatic effects of climate change, yet with more efficiency, the growth in population and services means that emissions will instead jump by 50%. Gates wants to reduce the population-based on this equation of fear. Bill Gates full name is William Henry Gates III, the "III" means order of the third (3rd). By converting the letters of his name into the ASCII values (which are used by computers and in the back of this book) you will get the following:

- Bill Gates III - 66+73+76+76+71+65+84+96+83+3 = 666

In 1798 Thomas Malthus an English cleric scholar, and an influential economist in the fields of political economy and demography was quoted as saying:

"Instead of recommending cleanliness to the poor, we should encourage contrary habits. In our towns we should make the streets narrower, crowd more people into the houses, and court the return of the plague. In the country, we should build our villages near stagnant pools, and particularly encourage settlements in all marshy and unwholesome situations. But above all, we should reprobate specific remedies for ravaging diseases; and those benevolent, but much mistaken men, who have thought they were doing a service to mankind by projecting schemes for the total extirpation of particular disorders. If by these and similar means the annual mortality were increased from 1 in 36 or 40, to 1 in 18 or 20, we might probably every one of us marry at the age of puberty, and yet few be absolutely starved".

I believe he like Boris and Gates was a social psychopath as he had 11 children but was afraid of over-population.

Let us look at the vaccine itself.

Here are the 4 steps that I saw going on around us in 2021:

1. Big Pharma funds Media
2. Media funds Fear
3. Fear funds so called Disease
4. So called Disease funds Big Pharma

Coronavirus

5 years ago, the word coronavirus was only in the vocabulary of virologists now, however, it is a household term spoken by many throughout the day, on news channels, radio networks and Internet search engines.

Coronavirus, however, spelt backwards is SURIVANOROC which seems to be an innocuous word but there is a warning here. If you take the language of Hindi and type in SURIVAN OR OC the resultant translation into English is SUNRISE AND EYE. So now we are back to the Eye of Horus and Novus Ordo Seclorum.

Taking the word CORONAVIRUS and looking at the lettering in accordance with the English language, there is only one word that is formed as an *anagram* from it which is CARNIVOROUS. We discuss zombies elsewhere, but this leads us down a sinister path.

If you word split CORONA VIRUS it gets even more interesting, reverse the words we get SURIV NOROC but if we move the A, then it becomes A NOROC SURVI which is a Romanian term which literally translates to GOOD LUCK SURVIVING so this makes it more frightening.

Let us look at the word in simple English Gematria.

C	=3
O	=15
R	=18
O	=15
N	=14
A+	=1+

6	66

The American elections in 2020 had interesting numbers in the campaign mixed with the word Corona.

Take 666 in the above calculation, then the year of the Biden campaign, 2020 and divide it by 666 = 30330 which was the text number chosen for his presidency campaign.

In another part of this book 3333 is mentioned but the above strap line text only has 3 x 3 so where is the 4th? Simple look at the red E in Biden, do you see the number spell now.

CIA

In 1959 on the 21st January, the CIA created something called the Corona program. This was a series of American strategic reconnaissance satellites produced and operated by the CIA Directorate of Science and Technology which was 6 years and 6 months in the planning. Run by the United States Air Force, the satellites were used for photographic surveillance of the Soviet Union (USSR), the People's Republic of China, and other American foes. It began in June 1959 and ended in May 1972. Interestingly in 2020 the first case of coronavirus in the United States was at 6 am 21st January exactly 61 years later to the day. If you take 6 years, 6 months at 6 am you get 666.

COVID

The tin foil hat conspiracy theorist wearers have turned into people with crowns of knowledge or Coron-et meaning crown. Corona is an anagram of raCoon which was the city from where the Umbrella Corporation created its virus in the video game franchise Resident Evil. When we look at the word covid but reverse the spell-ing we get DIVOC. There is no C in Hebrew, but the sound is represented by K and Q. The Hebrew verb lidbok means to Cleve, to cling. The Hebrew expression dybbuk me-ruah ha-rah refers to a wandering soul of the dead possessing a living body, a Jewish variant of possession by an evil spirit. The Arabic is Djinn. In the Hebrew dybbuk backwards in pronunciation is koovid. I think you can see a correlation here.

(C) = certificate
(O)f = of
(V) = Vaccination
(I) + (D) = Id-entification
1 = A
9 = I
Or Artificial Intelligence
Certificate Of Vaccination Identification Artificial Intelligence.

Vaccine

Vaccinae from the Latin word *vacca* – origin – cow. A Jew would call me as a non-Jew a *goyum* or cattle. A golden calf not to be worshipped. Throughout the vaccine rollout cattle is being used as a word only as a collective also known as a herd, let us look at the word. Thinning the herd, cull the herd, herd immunity, sheep herding, herd together, herd like, herd able, herded.

The opposite meaning of a word is called an antonym so *herd* as an opposite meaning includes *elite, elect, aristocracy, loner, member, single, entity, minority, clique, cabal, few.*

Herd can mean:

- Herd Human Experimental and Research Data Records (US NASA)
- Herd Human Exposure Research Division
- Herd -Evaluation and Reproductive Development
- Herd Health Environments Research and Design

The National Childhood Vaccine Injury act of 1986 (absolving VAX manufacturers of liability) was enacted 666 weeks before the Gates foundation was formed and 33 years 3 months and 3 weeks before the pandemic was declared. Here we have 666 and 333. 1986 is 1+9+8+9 = 24 and 2+4 =6.

The Jab

Prior to the year 1999 there existed a Swedish pharmaceutical company called Astra AB this company word lays in the Greek translation *A star*. At the end of 1999 it merged with a UK company called Zeneca Group plc.

In Hinduism astra (Sanskrit) refers to the word as a supernatural weapon, used by various deities giving them occult (hidden) powers to summon <u>Astra</u>. In Polish ze translated into English is <u>that</u>. We are rolling with words here with the chant of 3 but stay with me. If we again trace Neca its origin lays within Latin which is another spelling necare but remove the re and you are left with letters which directly translates to <u>Kills.</u>

Astra	Ze	Neca	re
Weapon	that	kills	

One of the companies providing PCR tests is called Ipsos Mori. Having a good comprehension of Latin, I saw straightaway that it translates into English as *Themselves Die* or *They Die*. So many more coincidences and occults.

Microsoft

In 2019 I was invited to do a lecture for the Alternative View Network exposing the following Cryptocurrency system patent for using body worn equipment and body activity data by Microsoft patent number W02020060606. The patent is so that the individual can transact or purchase products using a personal identification number (PIN) which would be worn but there is no reference to anything injected under the skin. While the 666 is obvious, there is something else I need to draw to your attention. The King James Bible mentioned the mark of the man being threescore and six. Using this old terminology, a score is 20. Now look at the patent number again but read backwards 60606002020. The 666 is obvious and the 2020 appears to be the year of application or 2020 linked to 666. Additionally, most home computers D/RAM (Random Access Memory) currently run at 2666 MHz Just another coincidence.

Luciferaise H1966
An oxidative enzyme Luciferin produces bioluminescence, a term created by Raphael Dubois who penned Telucifer in 1904. This word is derived from lucifer meaning light bearer and again I mentioned this in my 2019 lecture. Luminescence is something creatures can create naturally such as fireflies or jellyfish, but research had started to look at Quantum Dot delivery with certain exploratory vaccines. To date this has not been rolled out, but science is seriously looking into its potential for future use so that those who have been injected can be identified under ultraviolet light.

Dinoflagellate plankton also glows with this chemical so that predators will not eat them. Now look at the word dinoflagellate, do you see the word FLAG? You can also see EL in the word, a generic word for God that could be used for any God including Molech or Yahweh.

In 2017 Ted Talk, Moderna's Chief Medical Officer quoted *"and if you could actually introduce a line of code, or change a line of code, it turns out that has profound implications for everything from the flu to cancer"* when changing a line of code or introducing a line of code (referring to DNA), the code or DNA is then altered. The individual or subject has now had their Genome changed to what the scientists have coded. The individual or subject is no longer the creation of God but the creation of man, meaning the individual or subject could be the object of a patent. He went on to say that mRNA would tell the cells to code for the protein of the virus. This viral protein is foreign to the body. The individual's body is making foreign proteins for the immune system and opens it to attack. What is occurring is an autoimmune response is **A**uto**I**mmune **D**isea**S**e (AIDS). This all seems a little far-fetched, or does it? Let us go deeper.

Medicinal Symbols
In Greek mythology Hygieia was the daughter of and assistant to Asclepius, the God of healing and medicine. Hygieia's symbol was a bowl of potion surrounded by a snake demonstrating ancient wisdom. Depictions of Asclepius always has him with the staff as support which is surrounded with a snake ascending through its length.

The snake indicates the incisive and dissolving nature of Mercury. This reptile is the aspect of Mercury in its first state and the golden wand, it is the sulphate added to it or the philosophical Mercury and the caduceus as its symbol. The snakes spiralling around the staff alarmingly resemble the electromagnetic vortex. Mercury was known as the messenger of the gods.

The Bowl of Hygieia

The Caduceus below is a recognised symbol for the medical industry born out from the bowl of Hygieia and indicates where a pharmacist can be found on the High Street.

Hermes was said to be the messenger of the gods, but he was also according to mythology skilled in persuasion and bartering, and a trickster but more interestingly he considered the protector of merchants, travellers, **liars,** and thieves. Hermes is always depicted with the caduceus as a staff. The occult nature of this symbol for many of you reading will realise the symbol appears as the male organ on the groin in the androgynous depiction of Baphomet.

The Hermaphrodite of devil worship. The word Hermaphrodite was from Hermaphroditus the son of Hermes and Aphrodite and is the ancient Greek goddess of love and sex. Now exploring the Baphomet further, the many depictions of this mythical deity has its arms or hands marked with SOLVE and COAGULA, this is in fact a chemical process as a maxim or motto which means *dissolve and coagulate* or move implicitly to break something down so it can be rebuilt or changed. Baphomet with Solve, Coagula and Caduceus, but what does this demonstrate regarding the pharmaceutical industry?

As a sub note those of a religious propensity look at Deuteronomy 4:16 *"lest ye corrupt yourselves, and make you a graven image, the similitude of any figure, the likeness of male or female. The likeness of any beast that is on the earth, the likeness of any winged fowl that flieth in the air".* It continues *"The likeliness of anything that creepeth on the ground, the likeness of any fish that is in the waters under the earth".* Interestingly it was the fact that the Baphomet image had a stomach of fish scales which triggered my realisation to all the other aspects of this image. Are you starting to feel uncomfortable again? Follow me deeper and you will see what I will reveal to you.

101

As mentioned in a previous chapter George Orwell's 1984 as an interrogation room that dissidence go into and never come out and that within the Terminator movies model number 101 by Cyberdine Systems was the first bipedal robot sent back in time to destroy humanity. In the UK there is an entertainment show that puts ideas that are negative constructs into *Room 101*. While researching this number I found myself looking through American patents and found a court case dated 13th June 2013 *Molecular Pathology versus Myriad Genetics Inc.*

Under Patent law (to own an idea) you must not encroach 3 things:

1. Laws of nature - that which is natural cannot be owned.
2. Abstract ideas - a clear demonstration of what your patent does.
3. Natural phenomenon - a natural event cannot be owned. Chrysalis into butterfly.

A naturally occurring DNA segment is a product of nature and cannot be patent eligible because it is of nature, but C-DNA is patent eligible because it is not naturally occurring. According to this court case those who have had DNA alteration with C (change), the DNA can therefore legally be owned by the corporation that derived the original vaccine they have been injected with. Changed humans, therefore, will be owned as they are no longer considered human and therefore may not even have human rights.

Revelation 18:23

"The light of a lamp shall not shine in you anymore, and the voice of bridegroom and bride shall not be heard in you anymore. For your merchants were the great men of the earth, for BY YOUR SORCERY ALL THE NATIONS WERE DECEIVED".

The Greek Word for the Word sorcery / Sorceries in this verse and a couple other places in scripture is PHARMAKEIA. And is defined as follows:

SORCERIES: pharmakeia {far-mak-i'-ah}

1. The use or the administering of drugs
2. Poisoning
3. Sorcery, magical arts, often found in connection with idolatry and fostered by it
4. Metaph - the deceptions and seductions of idolatry

Therefore, the root word of *sorcery* is *Pharmacia* or pharmacists, so according to the above statement "AT THE END OF DAYS THE MERCHANTS OF THE EARTH WILL DECEIVE ALL NATIONS WITH PHARMACISTS".

The Baphomet Solve, Coagula and Caduceus as a medical symbol suddenly has sinister connotations. This is one aspect of this vaccination process that people have missed completely, only time will tell if it is an issue, but I genuinely believe it is.

70

On the 11th March 2021 as a timeline number President Biden *"aims to vaccinate 70% of adult Americans by July 4th"*. Nothing odd I hear you say but look at Reverse Full Reduction Gematria and the following words:

- Covid vaccine 6+3+5+9+5+5+8+6+6+9+4+4 = 70
- Independence Day 9+4+5+4+2+4+4+5+4+4+6+4+5+8+2 = 70
- Eleventh of March 4+6+4+5+4+4+7+1+3+3+5+8+9+6+1 = 70
- Coronavirus 6+3+9+3+4+8+5+9+9+6+8 = 70
- Low accountability 3+6+5+1+3+3+6+3+5+2+1+2+9+3+9+2+7 = 70

Order out of chaos - 12+9+23+22+9+12+6+7+12+21+24+19+26+12+8 = 222 another mystery school number. Interestingly from the Tuesday 11th September 2001, up to but not including Wednesday 11th March 2020, there was 6756 days or 18 years, 6 months, or 222 months. As these events unfolded on 11th March. Both the United States and the World Health Organisation declared that the coronavirus was a pandemic and major sporting events were cancelled from Friday 1st January 2021, to and including Thursday 11th March 2021, the total days between these two set dates in declaration was 70. But why is 70 relevant, we must dig even deeper?

The term *headquarters* really means 3 faces making 1 head. Multiple face worship exists throughout the world from the temples of Angkor Wat in Cambodia, Agni India, Isimud ancient Mesopotamia, it is endless, and I have visited these places for research. The duality of characters may not seem obvious but let us run through a few multi-faceted characters, Doctor Jekyll and Mr Hyde, Norman Bates and Mother, Tyler Durden, Revolver Ocelot, Francis Dolarhyde, Bruce Banner, Teddy Daniels, Batman even has an enemy called Two-Face or split personality. The many-faced god worship was even in Game of Thrones. The two masks are a symbol of the entertainment world. While Covid spelled backwards is DivoC. *Di* in Latin is "apart". The Greek translation, however, of voc is "to call" (voice) and *di* translates to twice, twins or two. So "twice call" or "twins call".

Janus Worship (RFR) - 8+8+4+6+8+4+3+9+8+1+9+2 = 70

Many people have never even heard of this worship. Today humans are all 2 faced, if you are unsure look at all the masked people, I refuse to wear one, so I am 1 face and the masked are another. Now let me introduce you to JANUS (Latin – Janus the J pronounced as Y) who in ancient Roman myths is the god of beginnings, transitions, gates, time, duality, doorways, passages, frames, and the *start and ending of things*. Associated with time, the horned god, and initiations. This gateway god was also worshipped for those travelling, trading, and shipping goods. All archways (access gates) give worship to JANUS. The word janitor derives from JANUS; he opens and locks doors or door keeper. His Roman number was 11. I questioned the doorway element, then I researched who was found hung on doorknobs in doorways and found the following:

- Robin Williams, Anthony Bourdain, Robert David Steele, Kate Spade, Chester Bennington, Chris Cornell, L'wren Scott, Aaron Swartz, Alexander McQueen, David Carradine, Michael Hutchence.

The patron saint of Freemasons is John the Baptist also known as John the Evangelist, and 2 Johns, bearing in mind John is a derivative of JANUS.

JANUS is the origin of the word January and the 1st day of each month at 1 am is a Janus worship time and his yearly celebration date was and still is the 9th January. Here we reveal 911 worship, the 9 is for Janus day and each 1 are the pillars of Freemasonry.

The old English tradition of opening the back and front door at the stroke of midnight on New Year's Eve to allow fresh air to pass through is a Janus ceremony, as is kissing for it involved with the meeting of 2 faces, a Janus greeting. The Roman military used to leave Rome under specific arches for luck, others were seen to be unlucky which still to this day exists with walking under ladders, the meeting of 2 pillars.

Janus continues with its link to pagan ceremonies:

- janua coeli - *janua* - Latin "the door", *coeli* - Latin "heaven door" (heaven's gate). The origin of the words for the summer solstice.
- janua inferni - *janua*- Latin "the door", *inferni* – Latin "inferno or hells door" (hells gate). The origin of the words for the winter solstice.

In an article published in 2009 by Timothy L Stinson, stating the possibilities of tracing mediaeval parchment through DNA analysis were produced for the use on several Greek manuscripts in an argument whether these parchments were made of animal or human skin. The test was carried out in 2006 using *polymerase chain reaction* which in planetary terms is now popularly known as the PCR test.

The name of the liquid handling PCR (Polymerase Chain Reaction) test machine that process Covid tests is called the G3 PCR workstation manufactured by PerkinElmer. This seems innocent enough but wait, I failed to mention its title.

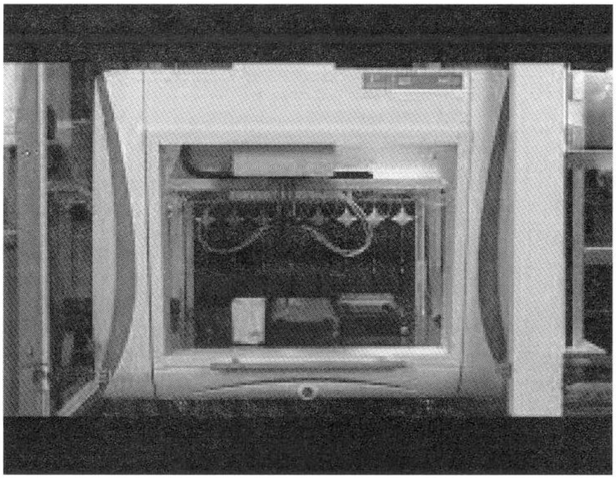

I introduce to you the JANUS G3 PCR. This device is set to cycle genetic sequencing and if the cycles is ramped up, they give a false positive. Starting to feel even more uncomfortable yet? Another fact is that in America The House of Representatives COVID-19 Testing, Reaching and Contacting Everyone (TRACE) Act is a document number 6666 or H.R.6666.

There is even a JANUS Island in Antarctica in the Palmer ARCH-ipelago at 64°47'S 64°06'W.

The JANUS Cosmological Model (JCM) which is based on the original theory released in 1977 by the French physicist Jean-Pierre Petit. It is a non-relativistic model based on Newtonian dynamics of two NRT enantiomorphic universes with opposite arrows of time. Known as the twin universe theory, so we gain a reference to 2 faces, 2 existences in our presence.

Even Saturn's Moons get a look in with Prometheus and Janus Saturn X being brothers.

Janus orbital period is 0.694660342 days.

The Plague

Jani Beg Controlled the Mongol Tatar Army from 1342 to 1357 which was called the Golden Horde. His army attacked the Crimean port of Kaffa in 1343 a north eastern province of Europe now under Russian control. Kaffa was a Genoese colony and included the shores on the Black Sea and was the epicentre for Europe's 13th century slave markets. The first biological attack on humanity was carried out by Jani during the first unsuccessful siege, he catapulted (trebuchet) bodies of the dead over the fortified walls and into Kaffa infecting the streets and water supplies. In 1347 another attack by Jani resulted in 4 ships leaving Kaffa and unbeknown to the sailors, deep within the ship's cargo were bodies infected with the plague. This was the source of the spread of the Great Plague of Europe. My research showed that Jani is the plural of Janus. The mystery continues because the ships ended up in Genoa (origin Janua meaning gate) from the word you guessed it...Janus. Janus was responsible for the death of half the population of Europe.

While the landing in Genoa caused the plague, take the words *The Genoa Plague* - 20+8+5+16+12+1+7+21+5+7+5+14+15+ 1 = 137 and 1+3+7 = 11.

The Europeans Plague - 20+8+5+5+21+18+15+16 +5+1+14+19+16+12+1+7+21+5 = 209 and 2+9 = 11. Kaffa was the start of great sufferance for humanity and as a place it is now called Feodosia which is a southern port in the Ukraine. It seems history is repeating itself.

Interestingly JANUS in reverse is SUNAJ in duality aspects it means – peaceful, systematic mind, mysterious. Sunjas is also mentioned in the movie *The New Jedi Order: The Final Prophecy*.

In the 1995 Bond movie *GoldenEye*, a previous MI6 operative (006) called Alec Trevelyan and played by Sean Bean adopted the identity JANUS. An anagram of Alec Trevelyan is *"reveal latency"*. Definition of *latency* is the state of existing but not yet developed or manifest concealment. Definition of *latency* in computing is the delayed transfer of sleeper data following an instruction for its transfer at a later planned date. Interestingly according to news reports Janus films are re-releasing *Eyes Wide Shut*. Janus Ja-sun or Jason was the mask wearer in *Halloween*, the killer with 2 faces.

Walking through a doorway resets the brain and is called the doorway effect. Ever wandered into another room and forgot why you were there. Well psychologists call it THE DOORWAY EFFECT. Human memories are episodic and not linear, which means they are broken into separate episodes. A new room is a separate episode in your mind's eye (hippocampus) and the memory can be jolted. People with Alzheimer's suffer these memory jolts which is why familiarity of a room with this disease is necessary to stay in the same 'episode'.

The God Gene

There is even a JANUS Gene in humans called Janus Kinase which controls cytokine, the male/female hormones. Duality of transgenderism. It is called the two-faced molecule which targets RNA DNA for gene editing.

Much conversation has been raised in the last 2 years about DNA RNA and cRNA with methods of vaccine delivery through this spike protein within our cells. In 2019, I returned as a speaker for another lecture for the Alternative View Network in which I mentioned something called the God Genome or Laminin. This sleeve of protection of our cells is an interesting aspect about the coronavirus. The injectables fool this Genome, called Laminin, which is the entire construct with the RNA alteration within our cells. Without Laminin we could not exist, and the vaccine needs to make it ineffective. Look at the shape of Laminin under an electron microscope.

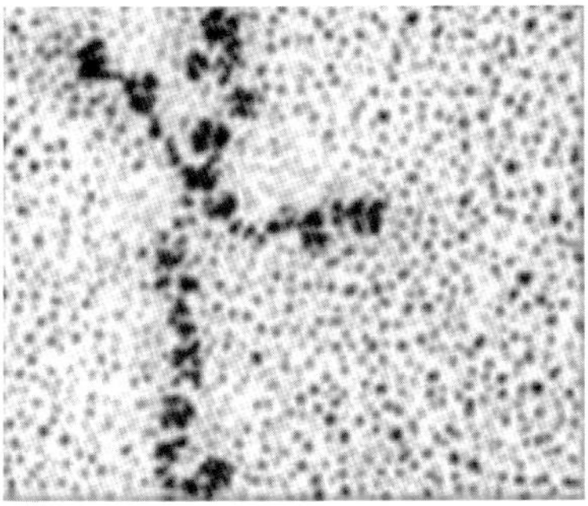

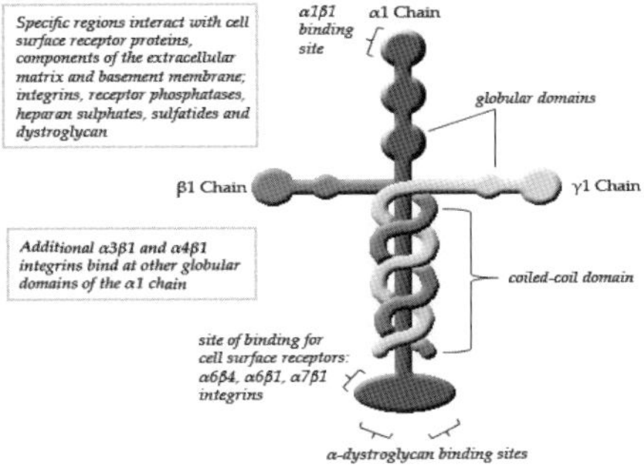

Behold, it is why it is called the *'God Gene'* for it is in the shape of a cross, a cross that must be fooled, avoided, and ignored for the RNA to change our DNA. I hope you can see my thinking on this regarding the injectable. It attacks our God Genome.

"Trust the science" is a term we have heard a lot of recently, but science is now a religion. Science has become a soundbite on the news. When I studied chemistry, it was all about method and protocols, it was the study of multiple papers and actual experiments. When I did this in the 80s it was about observing, realising a hypothesis, further observations based on more experiments and then producing a theory. This then led to testing the theory with scientific method. This data should then support your theory. Presently the data is now ignored so the theory becomes a religion through the mainstream media, science has just become a publishing house. Broadcasting has become spell casting. The media in my opinion is not in crisis, the media is the crisis.

DNA

Deoxyribo**N**ucleic **A**cid means:

Dioxi = God
Ribo = rabi master or lord
Nucleic - center
Acid - being on fire

DNA translates to *"god master and lord fire at the centre of my being"*.

DNA is the command centre of our bodies, when we are cut it gives commands to our cells to make new cells, to repair the damaged ones; to communicate between cells it uses a messenger or the M element of the M-rna with DNA.

Islamic Aspect

Firstly, let me explain some basic things to those who do not or have not studied Islamic teachings. Islam means "Submission to God". There are two main groups, Sunni and Shia who believe in one God, they follow the teachings of the Prophet Muhammad, who was Allah's last messenger on Earth. Muhammad is believed to have received the Muslim Holy Book directly from God and the book is called the Quran. For clarity, Muslims also believe Allah revealed holy books to other prophets who came before Muhammad and these books are called *REVEALED* books or *Kutub.* These books are mixed with other text, created by other people and therefore the only true word of the holy book is the Quran, the direct word of Allah. The book of all their books if you like, the source code.

Revealed books include the following:

Sahifah – The scrolls of Ibrahim, also known as the Suhuf.
Tawrat – This is known by the Jews as the Torah, revealed by God to the Prophet Musa which contains the 10 Commandments and sets out the "judgement of Allah" on non-believers.
Zabur – This contains poetic prayers and psalms believed to have been given to King Dawud, a form of update of the Christian Bible. Quran 4:163.

Injil – Believed to have been given to the Prophet Isa which is sometimes referred to as the Gospel of Jesus in Islam. Rather than believing Isa was the son of God as in Christianity, Muslims believe that the contents of this book revealed the coming of the Prophet Muhammad.

I should also point out the books **Hadith** and the **Sunnah** which are believed to contain the words and actions of Muhammad and give all Muslims advice and guidance on how to live their lives.

Allah Ta'ala states: *"They plot to extinguish the Noor of Allah with their mouths while Allah intends to complete His Noor even though the kaafiroon abhor it".* (Qur'an). Allah is mentioned 2699 times.

Isa (Jesus) is mentioned 25 times by name, 48 times in the third person and 35 times as a title. Interestingly Jesus is mentioned more times than the Prophet Muhammad in the holy Quran.

There is much rage within the Muslim world regarding the validity that vaccines contain Haram (unclean) constituents such as pig-gelatine, monkey kidney cells, horse blood, etc. This has caused fractious discussion. Most Sharia scholars, meanwhile, justify haram ingredients by invoking a concept known as 'transformation' (Istihala). This is essentially a recognition that things can change, which has been loosening things up ever since it was explored 1,200 years ago. The Quran's disapproval of wine did not rule out cooking with vinegar. Scholars have also reminded Muslims that out of necessity for public welfare emergencies can take priority over Islamic law's five goals (the maqasid al-shariah); which include the preservation of life.

Judaic Aspect

The Prophet Ezekiel (3:9–9) describes the weapons with which Gog and Magog fight in the terrible war. *"Then those who live in the towns of Israel will go out and use the weapons for fuel and burn them up. The small and large shields, the bows and arrows, the war clubs and spears"*. In the original Hebrew it mentions *"war clubs"* or *"javelins"* but in today's Hebrew language means *"hand sticks"*. While researching I came across the works of David Altschuler (1687–1769) and his interpretation which provided an interesting explanation. At the time he saw it as a deadly needle in the translation of the 300 years old prophesy. The rabbi wrote *"it is a long wooden stick with an iron needle on its head that kills people"*. It was the word needle and not tip of the spear which I found interesting. **Pfizer** spelt backwards is Rezifp which in Hebrew is pronounced Resheph. In ancient Canaanite this is a noun interpreted as "flame, lightning" but also "burning fever, plague, pestilence".

Other Belief Systems

Hindu, Sikh, and Buddhist texts that I have read intimate that the concept of preventative medicine is not part of a man's existence during his life and intervening in any way, including surgery or blood transfusion should be rejected as it is "gods will" if you like. The background basis of these teachings seems to be based in the belief, not of religion or tradition, because the belief is that it interferes with the system of reincarnation (connection to god/ the creator) or spiritual advancement. It is this reason that the injectables are being refused within their belief system. Hindu followers believe it disconnects man to god.

666 and DNA

Carbon 12 on the periodic table is one of the 5 elements in the human DNA Genome, it is composed of 6 protons, 6 electrons and 6 neutrons, giving you 666. "It's star stuff" (Carl Sagan) due to its abundance in the triple (3) alpha process in the make-up of carbon. It is the basis of our physical matrix. Physically the human body is linked to 666. Could this be the mark of man, not the beast?

Carbon derived from the word *Cabo* - Latin for "coal or charcoal". In Isaiah 6:6, *"Then one of the seraphim flew to me with a live coal in his hand, which he had taken with tongs from the altar"*. The name seraphim meaning "the burning ones", of the first hierarchy of angels who were to become Demons. Beelzebub (Baal-ze-Bub) was the second in command below Lucifer (dealt in another chapter). John the Baptist mentioned the coming Antichrist and the number of the beast *"but him that has understanding count the number of the beast for it is the number of man and his number is 600, 3 score and six"*.

In English Ordinal when you add the words Carbon in the Periodic Table Equals - 3+1+18+2+15+14+9+14+20+8+5+16+5+18+9+15+4+9+3+20+1+2+12+5+19 = 322 another esoteric number appearance. With carbon, humans are cosmically linked to the numbers 666.

Shakespeare and the Zodiac

While writing this chapter I was contacted by a man called Alan, so this paragraph is down to him, thank-you Mr Ayre for sending me down this rabbit hole. The first jabs in the UK were administered at 6.31 am on the 8.12.2020 in Coventry. William Shakespeare was the first male recipient, and the other player was a nurse who administered the vaccine called Ms May Parsons. The ruling number of the day was 6 via the date (8+1+2+2+0+2+0 = 15 = 6), in the planetary position of Venus, the ruler of veins.
The day was = 6
The time was = 6
May Parsons (EO) 13+1+25+16+1+18+19+15+14+19 = 141, 1+4+1 = 6

Upon that day Venus (6th Planet) was debilitated in Scorpio and we all know Scorpions are poisonous with a needle stab. COVID-19 equals DVICO19, Templars were founded in 1119, the Sun has 11-year cycles, and the Moon has 19-year cycles, so here the 1119 appears.

William Shakespeare was the first male to receive the Pfizer-BioNtech COVID-19 vaccine but here are more interesting numbers in English Ordinal:

- William Bill Shakespeare - 23+9+12+12+9+1+13+2+9+12+12+19+8+1+11+5+19+16+5+1+18+5 = 222
- Wuhan coronavirus - 23+21+8+1+14+3+15+18+15+14+1+22+9+18+21+19 = 222
- Mandatory vaccination - 13+1+14+4+1+20+15+18+25+22+1+3+3+9+14+1+20+9+15+14 = 222
- September eleven attack = 222
- Event 201 was a coronavirus pandemic drill run by the World Economic Forum. Event two zero one = 222
- World Economic Forum - 23+15+18+12+4+5+3+15+14+15+13+9+3+6+15+18+21+13 = 222

- The Great Conjunction - 20+8+5+7+18+5+1+20+3+15+14+10+21+14+3+20+9+15+14 = 222
- The first ever purpose made vaccine for smallpox was in 1796 and rolled out in 1798, bring that date to 2020 and the difference is exactly 222 years
- The infamous date of 9/11/2001 and 3/11/2020 the date that everything changed with the virus has a correlation for it equals 222 months

The Pfizer jab study will not be complete until 27th January 2023, Moderna trials end in December 2023 and AstraZeneca trials will reach completion in February 2023, nothing appears in relation to the months but look at years. These trials are being carried out over a three-year period, add them up from the year they started:

2021
2022
2023 +

6066 just another coincidence

Consider the year 2022, this number breaks down to 3 instances of the number 2 or 3x2 so according to the mystery schools what is planned for March = 3 in 32 or 322?

Gematria
If we play with words and things start to look a little interesting:

56
The Californian state was locked down in March 2020 by Governor Newsom. He stated that in 8 weeks (56 days exactly) 56% of California would get coronavirus, an extremely specific prediction I would say and this made me review the numbers with the following word associations:

- Coronavirus (FR) - 3+6+9+6+5+1+4+9+9+3+1 = 56
- Covid vaccine (SR) - 3+6+4+9+4+4+1+3+3+9+5+5 = 56
- Anthony Fauci (SR) - 1+5+2+8+6+5+7+6+1+3+3+9 = 56
- Gavin Newsom (RFR) - 2+8+5+9+4+4+4+4+8 +3+5 = 56
- Washington DC (FR) - 5+1+1+8+9+5+7+2+6+5+4+3 = 56
- Society of Jesus (FR) - 1+6+3+9+5+2+7+6+6+1+5+1+3+1 = 56 (both Newsom and Fauci are Jesuit educated)
- Mind Control (SR) - 4+9+5+4+3+6+5+2+9+6+3 = 56
- All Seeing Eye (FR) - 1+3+3+1+5+5+9+5+7+5+7+5 = 56
- Unemployment (FR) - 3+5+5+4+7+3+6+7+4+5+5+2 = 56
- Toilet Paper (FR) - 2+6+9+3+5+2+7+1+7+5+9 = 56 (this one did make me chuckle)

Boris Johnson at the age of 56 released a 56-page Winter Plan for coronavirus in 2020 before announcing the new and more infectious UK strain, which again was reported to be 56% more infectious. In the same year, the Jesuit Pope wrote *Fratelli Tutti* his encyclical regarding what to do about coronavirus as well as climate change; now consider the following:

- Fratelli Tutti (FR) - 6+9+1+2+5+3+3+9+2+3+2+2+9 = 56

- Climate Change (FR) - 3+3+9+4+1+2+5+3+8+1+5+7+5 = 56
- All this under the guise of the Paris Climate Accords and 'Paris, France' adds up to a convenient 56

Why is 56 important? Look deeper from a mystery school perspective at the number 5+6 = 11.

80

- Pfizer (EO) - 16+6+9+26+5+18 = 80
- Baphomet (EO) - 2+1+16+8+50+13+5+20 = 80
- The Beast (EO) - 20+8+5+2+5+1+19+20 = 80

333

- Corona is in the vaccine (RO) - 24+12+9+12+13+26+80+80+80+13+7+19+22+5+26+24+24+18+30+22 = 333
- Corona vaccine deaths (RO) = 333
- Moderna Covid Vaccine (RO) = 333

English Sumerian Gematria and 666

- RFID Scanner 108+36+54+24+114+18+6+84+84+30+108 = 666
- Satan's RFID 114+6+120+6+84+114+108+36 = 666
- Vaccination 132+6+18+8+54+84+6+12+54+90+84 = 666
- Mark of beast 78+6+108+66+90+36+12+30+6+114+120 = 666
- A Covid Vaccine 6+18+90+132+54+24+132+6+18+18+54+84+30 = 666
- Bio Implant 12+54+90+54+78+96+17+6+84+120 = 666
- Mandatory 78+6+84+24+6+120+90+108+150 = 666
- Receive a Mark 108+30+18+30+54+132+30+6+78+6+108+66 = 666
- The Hand or Head 120+48+3+48+3+6+24+90+108+48+6+84+24 = 666
- A Satanic Mark 6+114+6+120+6+84+54+18+78+6+108+66 = 666
- Evil Masks 30+132+54+72+78+6+114+66+114 = 666
- The Face Masks 120+48+3+36+6+18+30+78+6+114+ 66 +114 = 666
- People Sin 96+30+90+96+72+30+114+54+84 = 666
- Monetary 78+90+84+30+120+6+108+150 = 666
- Lucifer Hell 72+126+18+54+36+30+108+48+30+72+72 = 666
- Lucifer Hades 72+126+18+54+36+30+108+46+6+24+30+114 = 666
- Corrupt 18+90+108+108+126+96+120 = 666

The Health Minister of the UK, Mr Sajid Javid announced there is a new South African strain number B.1.1.529 and it was given the name Omicron, an acronym for moronic. In the announcement this variant has 30 mutations, twice as many as the Delta (Indian) variant but look at the numbers 1+1+5+2+9 = 18 and 18÷3 = 6. They do like their numbers. Additionally, Omicron was announced to have hit the UK on 27th November 2001. Bearing in mind Coronavirus was first announced on 31st January 2020, exactly to the day the calculation is 666 days. Then consider that 31st January 2020 was the 31st day of the year in the Gregorian calendar which means there were 335 days remaining until the end of the year and if you add these 3 numbers, they reduce to 11.

The 1st lockdown in Ireland was on 27/3/20, restrictions ended on 22/1/22 and the total days of restrictions is 666. The UK 1st lockdown was 23/3/2020 and the announcement of the end of restrictions was 19/1/22. The total days of restrictions was 666.

Other interesting numbers are:
Bill Gates - 2+9+12+12+7+1+20+5+90 = 87 in (FR) 33
Person - 16+5+18+19+15+14 = 87 in (FR) 33

Genetic Code
Before 2020 only 12 MRNA vaccines were made for human trials but more importantly were not approved, then came the coronavirus. It is argued that the warp speed development of this new innovative intervention was driven by urgency created by the demands of governments around the world. Since then, the human trials on gene manipulation have been rolled out and humanity is over halfway through its test. I for one have refused to join this medical trial and my analogy to people is: *"tomorrow we're going skydiving but before we jump out of a perfectly good aircraft, you'll be presented with five parachute two of those parachutes Will not work and you will fall to your death... are you coming"*? To this answer I usually get *"that would be insane"* and my retort is always *"then why take part in the experiment"*. Revelation 16:2 *"And the first went and poured out his vile upon the Earth; and there fell a noisome and grievous sore upon the men which had the mark of the beast, and upon them which worshipped his image"*. Look at all the rashes and boils that people are suffering because of the vaccination.

To add to this position, I have spent many weeks researching all avenues regarding this experimentation, and I have come to the conclusion that viruses may not even exist. I believe peoples immune systems will be compromised after this intervention and the subsequent injuries and deaths of humanity will ensue with incalculable numbers. Some may scoff at that statement. I usually get the aggression toward me starting with "you're just a builder". but I stand my ground and leave time to tell the tale, and history to judge me. I started this chapter with the Georgia Guidestones, and I believe there is a direct correlation with the first commandment upon those stones and the mandates or offer we are suffering.

The Future
As the world so called pandemic rolls on there will be 2 possible outcomes, the 1st being a dystopian future, cashless transactions, tracking systems and authoritarian control. The other will be a world where people realised what has been done to them and answers will be demanded of those who were in control. Politics is a communication system of control and when politicians themselves are held to account they use 5 excuses from their play book:

1. There is a perfectly satisfactory explanation for everything, security prevents its disclosure.
2. It has only gone wrong because of heavy cuts in staff and projects which have stretched supervisory resources beyond the limit.

3. It was a worthwhile experiment now abandoned, but not before it provided much valuable data and considerable employment.
4. It occurred before certain key facts were known and could not happen again. Lessons have been learned so this will never happen again.
5. It was an unfortunate lapse by an individual now being dealt with under internal disciplinary procedures.

Numbers 3 and 4 will be used if they are ever held to account.

Lastly the mathematic equation to watch out for in the coming years is this:

- BxCxD=AHH which is the Codex that will be used to Hack the Human Genome. B = Biological Knowledge, C = Computing, D = Data and AHH = Ability to Hack Humans. Time will tell and history will judge my prediction.

Those of you that read *stopper 1* will understand the difference between the word *person* and *man/woman*, it is a duality of our existence between those in legalese and those in common law. So, in this chapter those of you who are deep thinkers now know what Janus worship is and its deceptions of being two identities, both the masked and the unmasked, I therefore leave you with this tile for consideration.

THE TWO

WORLDS

OF

CREATIONS

NATURAL		ARTIFICIAL	
NATURE/GOD Nature/God is over Man/Woman >	MAN/WOMAN Man/Woman is over Government >	GOVERNMENT Government is over Corporation >	CORPORATION Corporation is over **nothing**
Natural Man/Woman Human Being, Living Factual Entity, Creditor, Flesh & Blood, Sentient, With Spirit *status naturales* *John: Doe*		Artificial Person Legal Person, Dead Fictional Entity, Debtor, Creature of the 'Law', Without Senses *universitates bonorum* *JOHN DOE*	

DOMINIONS

THE UNIVERSE PLANET, EARTH, FREEDOM	THE EMPIRE REALM, STATE, KINGDOM
Man/Woman/People	King/Queen/Crown
Innate Human Rights Unalienable Rights, ***un-a-lien-able*** rights Endowed by Creation	'Divine Right of Papacy' to the World & Souls, *unam sanctam* 1302 & 'Divine Right of Kings'
Universal Birth Rights	Proclaimed Birth Rights
Govern by *Right* TRUE REPUBLIC / DEMOCRACY By the People, Of the People, For the People, Equal Representation and Protection in Law (People's *s*Elections)	Rule by *Claim* MONOTHEISM, MILITARISM, MONEISM, MONARCHISM, > Aristocracy, Oligarchy, Plutocracy (Family Bloodlines)

P.S An interesting statistic, the chance of the Earth being hit by an asteroid is 0.046% but when I worked out the chances of dying of so called Covid it is 0.026%. With this analysis I suggest that those fools wearing a mask should in fact buy a crash helmet first.

Chapter 6 - Roulette, Snakes and Squid Game

Games

Within Revelation 13:18 *"This calls for wisdom: let anyone with understanding 'reckon' the number of the beast, for it is the number of a person. Its number is six hundred and sixty-six".* Reckon- origin from the Greek is *"to count".* The ability for you to calculate is considered essential in revealing the depths in language. Most games are based on numbers, here are a few and the occult that sits behind them.

Roulette

As a circular spinning dish, it uses a 1 to 36 chance of it landing within the number you choose, with 36 being a triangle number. This game is sometimes called the "Devil's Wheel" or the "Game of 666". It is said that François Blanc had done a deal with the devil to introduce the single 0 style roulette wheel and therefore make him wealthy.

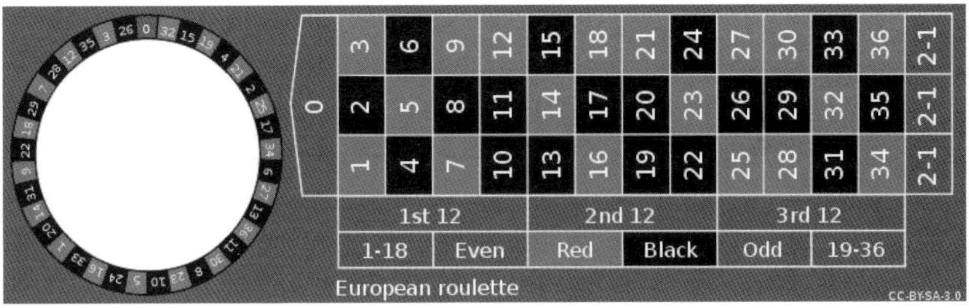

Start with adding all the numbers on the wheel itself - 1+2+3+4+5+6+7+8+9+10+11+12+13+14+15+16+17+18+19+20+21+22+23+24+25+26+27+28 +29+30+31+32+33+34+35+36 = 666

If you look at the table layout and add the verticals, something interesting appears. Take the first vertical 3+2+1 = 6, then moving to the right 6+5+4 = 15 but 1+5 = 6. This codex runs throughout all the 12 verticals, ending with 34+35+36 = 105, 1+5 = 6. Now look at the diagonals starting with the bottom right-hand number, the 34 red and work up the reds 34+32+30 = 96 and 9+6 equals 15 and 1+5 = 6, frighteningly this happens throughout all the diagonals as well.

If you look at the horizontals, there are 3 lines of 12 numbers with 1+2 = 3, so where are the other 3s hiding? You can bet 3 ways, on a single number, odds, or evens or black and white, so we find the 333. If you look at the red and black squares, they total 12 and 1+2 = 3 so we find 333 and 3.

James Bond has been in 2 films called *Casino Royale*, the first with David Niven (1967) and the second with Daniel Craig (2006). His gambling exploits with Baccarat and Poker were deep seated and his favourite gambling number call was 17. If you look closer 007+1+7 = 15 and 1+5 = 6. Created in 1842 this creates another coincidence 1+8+4+2 + = 15 and 1+5 = 6.

In *Casino Royale* there are a few paragraphs about his exploits at playing roulette. The secret agent uses a betting system to overcome the odds against him, with the system being called the "Labouchère system", nowadays it is referred to as the "James Bond Roulette Strategy".

- $\sum_{n=1}^{666} 2n(-1)^n = 666$
  interestingly this is the equation to work out the mathematics on a roulette wheel

Dice and 7

Put this book down and go and find a dice...off you go...I keep telling you to stop reading this book...have you found one?

The earliest dice were made from knucklebones of sheep or humans and were considered as a method of communicating with the gods, spirits, and the dead, which is known as *Astragali* and *Pessomancy*. Largely used to predict the future in divination and cleromancy, *"will of the Gods"*. Ubiquitous throughout all cultures and civilisations, dice are as old as history itself. The earliest form of dice was found by the British Archaeologist Sir Leonard Wooley in 1923 in Ur Iraq. Pyramidal in shape and 4 sided, they are dated from early 2600 BC.

Ancient Roman houses had dice rooms to predict the god's favour. When General Julius Caesar led his army across the Rubicon River to invade Rome in 49 BC in his commence to power, he proclaimed *"alea iacta est"*- "the die is cast". This is where we get the terminology, we have crossed the Rubicon, meaning there is no turning back. Loading dice is adding weights by various methods to creating a heavier side, birthing the term "the dice are loaded". As dice were originally made of bones, the term "rolling your bones" is slang when playing with dice used in a gaming house.

The 6-sided dice is mathematically based on the number 7. Consisting of 6 sides, each side holds a number from 1 to 6. So, how comes 7? If you throw the die and it shows a 6 face up then its opposite face down is 1, which adds to 7. Throw a 5 and its opposite face is 2, adding again to 7.

Time and the number 7 have a direct relationship with dice. Look at a clock face and choose a number then look at its opposite number across the clock face. For example, the number 10 is directly opposite 4. Take the lowest of these 2 opposing numbers and subtract it from the larger for example 10-4 = 6 , add 1 and you will get 7. Do this with all numbers on the clock face, remembering to add the 1 and as in the dice you will always get 7.

The 6 numbers on the dice are 1+2+3+4+5+6 = 21 and 21÷3 = 7 or 3 sets of 7 are 7+7+7 = 21.

Russian Roulette

This is a game which involves chance and a handgun (in the manufactured format of a revolver). An automatic holds the bullet rounds in the handle and are far more modern in design than a revolver which holds them in a cylinder. With this design there are a total of 6 chambers available for 6 rounds or bullets. Using only 1 bullet and dropping it into a chamber, that in turn folds into the main housing of the weapon you are ready to play. Spinning the chamber into a random position and then pulling the trigger against your head or in your mouth is truly chancing your luck.

There is a 1 in 3 chance you will die and a 66.6% chance you will live. Another game full of 6s.

Occult Board Games
Over the last 20 years board games with occultic symbolism have exponentially increased with devil worship, evil worship; summoning such the original as the Ouija board game which again are on the increase. These include:

- The Mayan Calendar, Alchymia, Apocalypse, Arcanum, Arkham Horror, The Game of Prediction, Astro Magic, Battle for Souls, Black Magic Ritual Kit, Cave Evil, Cosmic Karma, Danse Macabre, Dante's Inferno, Dark Cults, Day & Night, Demons, Do You Worship Cthulhu, Doctor Faust, Freimaurerei, The HellGame, HellRail, Magdar, Chaos Progenitus, Demonlord, Demono, Demonworld, Auf Teufel Komm Raus, Hol's der Teufel, In Teufels Küche, Necromancer, Diabolo, Oh Hell!.

Steve Jackson
In 1995 Mr Jackson created a card and board game called the Illuminati. The game is built on conspiracy theory but some of the cards seem a little more coincidental than conspiracy. Other games he created include Munchkin, Zombie Dice and Cthulhu Dice. There are great discussions whether his games are predicted plans, prophesy, coincidental or about manifestation.

When we look at the background of Jackson's games there is no doubt, he is an occultist. On the morning of 1st March 1990, a force of armed US Secret Service Agents supported by the Austin Police Department occupied the offices S J Games in the search for all the computer equipment. The raid zeroed in on fraud 'supposedly' committed by the company. The case went to court but was thrown out and S J Games were awarded $50,000 plus $250,000 for legal fees. The president at the time was George W Bush and why was he so driven to try to ban this game?

The Illuminati card game has had multiple releases since the 1st edition which include the Mutual Assured Distraction, Bavarian Fire Drill, Y2K, New World Order, and each with expansion packs of cards. The basis of the game is like Monopoly in that the winner is the one that controls everything. The idea is to get money through terrorising the planet. The first step in the game is to join a network of 1 of the following groups:

1. The Bavarian Illuminati
2. The UFOs
3. The Servants of Cthulhu
4. The Society of Assassins
5. The Gnomes of Zurich
6. The Network
7. The Bermuda Triangle
8. The Discordian Society

These groups fall under 10 political alignments: *government, communist, liberal, conservative, peaceful, violent, straight, weird, criminal*, and *fanatic.*

As you progress through the game using 2x6 sided dice, the idea is to take control of other 7 secret society groups and use them to do your bidding. Here are some numbers for you:

- Release date - 1982
- From the original books of Illuminatus! Trilogy (3) by Robert Anton Wilson and Robert Shea published in 1975
- The original number of cards was 110

People who materialise rabbits from hats, join mental rings, or saw women in half have moved into the digital age with a game called *Conspiracy* by the Magic Gathering. It involves puzzles being put out into the magic community to solve puzzles. Solve the puzzles get the cards, creating a hundred-thousand-dollar industry. A cartel has been created driving up prices in this collectables market.

- 54 that appears in other games
- Number of cards in a deck
- Number of squares on a Rubik's Cube
- Jenga block game for children has 54 pieces

The 4-11-44 Phrase

Music and gambling have a direct relationship with these numbers. In stopper 1 I wrote about the Devil's Chord but let us investigate the numbers themselves. While this number has no meaning now, in the late 19th and early 20th century it was a term widely used. In the US illegal gambling using a wheel of fortune machine of 1-78 was extensively played and a 3-card section was called "doing a gig", the same term used by musicians for playing a set at an appearance. The term "gig" replaced it for a range of songs the band gambled on the fans liking that evening. In the year 1895 Charles August-Fey of Bavarian birth and living in San Francisco as a mechanic invented the world's first coin operated slot machine as a bar room entertainment device, his second machine was called the "4-11-44". In 1898 automated coin pay-out version was created with half the coins stored for the bar owner and the remainder was for the player and was called the Liberty Bell. The later invention had 3-reel spinners with symbols like today's slot machine. In the US there is 1 slot machine for every 357 citizens 3+5+7 = 15 = 1+5 = 6 and with those odds they really are "One Arm Bandits".

Cards

When you shuffle a deck of cards, do you realise the mathematical insanity that is involved for you to shuffle the same deck in your lifetime? The chance of you doing this twice is a factorial calculation, which means you start with 52 cards then multiply by 52 and again by 51 and gain by 50 and you continue throughout the deck. This works out to 8×10x67 ways to sort a deck of cards. That is an 8 followed by 67 zeros or
80000000000000000000000000000000000000000000000000000000000000000000.
If we put this in the perspective of time, if someone could shuffle a deck of cards every second of the history of time, they would not get 1 billionth of the way to finding a repeat of the same

deal of cards. The chance of shuffling the deck of cards into perfect order spades, hearts, diamonds, and clubs is 1 in 10 to the power of 68 (followed by 68 zeros).
If each person on the planet shuffles 1 deck of cards each second for the age of the known universe in time, they will be a 1 in 1 trillion, trillion, trillion chance of 2 decks matching.

There is only 7×10^{22} stars in the known galaxy, and the combinations of playing cards is too cumbersome to put into words, so here is the number: 80,658,175,170,943,878,571,660,636,856,403,766,975,289,505,440,883,277,824,000,000,000,000. Because a deck of cards has 52 unique cards. The combinations of cards are greater than the number of observable stars in the galaxy. The numbers in shuffling cards are literally astronomical.

Chessboard
Consist of 64 squares in an 8x8 pattern, interestingly 64 is a square number and this is the reason the chequerboard square is used on Freemasonry floors and as is the question "Are you on the square". The allegory is "are you ready to play a game". Freemasonry levels are called degrees and the better the player, the higher degree of skill. The word checkmate is to signify the end of the game by successfully holding the king in a trap. The word derived from the Persian word *Shah* (King) and the Arabic word *Anah Myit* (he is dead).

The 64 squares are divided into black and white, both being 32 each. The number 64 is a multiple of the fundamental cyclic number 25920, the exact measurement of the precession of the Equinoxes in a long year. While this board and the Masonic tile floor have a relationship, be aware that the king (most important piece) represents the Sun and the queen (most powerful piece) the Moon. The middle of the board represents the centre of the universe in astrological terms. The linear outer edge consists of $4\times8 = 32$ squares representing days of a monthly lunar cycle with each side representing fire, earth, water, and air. An octagon can fit perfectly upon the board and the inner angles add up to 1080° which is the radius of the Moon in miles including surface rise and fall, otherwise it would be 1,079.6.

Total pieces add up to 32, with each 1 granted a set range of movement. There are 20 possible opening moves, 400 possible second moves and 20,000 possible third moves; these numbers increase exponentially as the game proceeds. There are only 2 numbers that repeat, a 3-digit pattern and these are 37 and 27.

- $1\div37 = 0.027027027$ or 270-270-270 etc
- $1\div27 = 0.037037037$ or 370-370-370 etc

What if you take $37+27 = 64$, the total square is on the board a mirror of themselves $37\times27 = 999$.

Snakes and Ladders
This board consists of 100 squares in a 10x10 grid. Originally from 19th century India, it is about the moral message of virtues and vices. Originally it was called Gyan Chaupar (Game of Knowledge) and had ladders and chutes or slides. This took on a spiritual meaning for children as a game. The design always has more snakes and ladders, and the occult is that life path is filled with more bad decisions than good. The game reached England in 1892, with returning military personnel bringing it back from India.

Snakes and Ladders is a game of serpents and Jacob's ladder to God and the serpent of god who deceived Adam and Eve. Originally there were 9 ladders and 10 snakes. Snakes and ladders are about the Garden of Eden. Too many snakes on the top row would mean the cycle would take too long and the game becomes frustrating. Too many ladders and it would end too fast; the most optimal layout of making the game balanced is as follows:

Ladder	From	To
1	1	38
2	4	14
3	9	31
4	21	42
5	28	84
6	36	44
7	51	67
8	71	91
9	80	99

Snake	From	To
1	98	78
2	95	75
3	93	73
4	87	24
5	64	60
6	62	19
7	56	53
8	49	11
9	48	26
10	16	6

In the Demons part of this book, I mention an old black-and-white movie called *Night of the Demon.* In the movie, the villain *Karswell* who is the demonologist is hosting a Halloween party where 2 young children are playing the game. Consider this game when you read that chapter. You can now see this game as a demonstration that life is filled with opportunities and threats, wins and losses, what we desire and most fear on our journey takes on a different context. After all this is about 2 humans in a race or a Human Race.

People call the game "Snakes and Ladders", but I have always preferred "Adders and Ladders". The lines in the AC/DC song Sin City "*Ladders and snake Ladders give Snakes take Rich man, poor man Beggar man, thief Ain't got a hope in hell That's my belief*" always makes me smile and was never lost on me.

Boxing Ring

Boxing rings are squares, but the origin comes from ancient fighting practices such as sumo which is a ring.

The first relics of ring fighting rules that is documented is from the British Pugilists' Association of 1838, the ring or square was specified 24 feet along each side then take 2+4 = 6. The cable rope surrounding it was 6 feet high. Take the four sides, 24×4 = 96 then 9+6 = 15 then 1+5 = 6. Once again, we have 666. Many years ago, I was given an instruction to redecorate the exterior of 109 Piccadilly, unbeknownst to me when I applied for listed building consent the information returned was that it was originally the house of the 9th Marcus of Queensberry who created the Gentleman's rules of boxing. The rounds were of 3-minute intervals. There would be a total of 12 rounds 1+2 = 3. If a man fell to his knee and was counted out more than 3 times the match was over, called a **T**echnical **K**nock **O**ut (TKO). Then let us not forget that are 3 judges making score cards, so we really have 3333.

Squid Game - a critique of capitalism

These next few paragraphs could have been placed in the movie section of this book; however, the occultism of the game being played found itself more appropriate for this chapter.

Last year on Netflix the *Squid Game* series captured people's imagination and Netflix revealed that more than 142 million households viewed it in the first 28 days on its platform. The series is a simple concept, the wealthy having fun with those in debt who must play a game of life. The WEALTHY VIP's wore costumes as if they were at a Rothchild's party with masks made of gold depicting decadence and mankind's animalistic nature and represent the 7 deadly sins as follows:

The bear	Sloth
The eagle	Pride
The tiger	Wrath
The deer	Envy
The buffalo	Gluttony
The owl	Lust
The lion	Pride

The SUPERVISOR of the game was ex-police, who is in the pocket of those who commanded the game. The individuals that controlled the game had facemasks of triangles, squares, and circles; the same symbols on a PlayStation CONTROLLER. Circles were the lowest rank and can only speak when spoken to. Triangles were the middle run of the hierarchy and act as soldiers. Squares were affectively the bosses within the working group and keep their fellow circles and triangles in check and are in charge. The dead or losers were placed in black boxes with ribbons as gifts to the gods, their organ removal paid for the prize money to the last debt slave standing.

The 6 games were as follows:

- 1: Red Light, Green Light
- 2: Dalgona/Ppopgi
- 3: Tug Of War
- 4: Marbles
- 5: The Glass Tile Game
- Final Round: Squid Game

The *Last Supper* is held in a room with a triangular table setting above a chequerboard floor with 2 pillars of light referring to Boaz and Jachin, the Pillars of Freemasonry.

The plot twist at the end with the main protagonist being invited to an apartment on 24th December at 11:30 pm in the Sky Buildings 7th floor. Upon arrival he meets Player 001 laying in his deathbed with Gi-hun questioning why such wealthy people would create such a horrific game. His reply was "*if you have too much money, it doesn't matter what you buy, eat or drink, it all gets boring. All my clients eventually started saying the same things when we talked. Everybody felt there was no joy in their lives anymore. And so, we decided to get together and started asking what we all could do to finally have some fun?"* A bunch of elites with too much money having done everything in life started to enjoy the death of others.

The players numbers are quite interesting, the man who created it was number 001, the villain was number 101 and the winner was number 456. Player 218 a corporate failure, 199 has not been paid by his employer for 9 months. These numbers according to the script were the order in which they were recruited to play the game. The game is rigged by the elite and soon has them turning on each other describing what happens in life by those who have no money.

Now some numbers for you:

- 9 episodes
- 456 - 4+5+6 = 15 - 1+5 = 6
- 218 - 2+1+8 = 11
- 101 discussed elsewhere
- 001 the beginning

The Squid Game telephone number turned out to be real and a man was inundated with telephone calls to his home with people wanting to play the game. At the time of writing this chapter he had so far been offered $4,800 for the number, with the purchaser wishing to remain anonymous and their intended use for it in the future is also unclear.

I felt extremely uncomfortable watching it and my reasoning is this; if you watched the series with the millions of others, consider your enjoyment may be the same as the elite watching people die, while eating popcorn. If the game was real and televised ask yourself, would you tune in?

Chapter 7 - The Sun, Sun Dials and Rolex – other products are available

Mankind has built structures throughout our planet. First it was for shelter, then places to worship and communal places to gather in halls of kings, taverns to talk and castles for safety. From a European perspective, as populated areas grew villages had public houses, towns had churches and cities had cathedrals. The powerful have always reached to the skies to demonstrate power over the populous. From the Tower of Babel, the spires of churches all ascending to the sky. Even today the powerful continue this architectural demonstration with towers, penthouses, skyscrapers, and bank headquarters. All reach for the sky and to God.

When the powerful become wealthy, the demonstration of their understanding of what they consider success reflects more about these individual's narcissism and intention in a physical representation of what they leave behind. In England wealthy individuals who owned large properties and swaths of land often built follies, which were structures of extravagance primarily for decoration. The difference between an insane man and eccentric one, I would argue is money. Over time these man-made formations started to have significance in the form of messages being laid out to those who were educated and aware. The aesthetic nature started to leak into other more complex messaging. The worship, delight and light heartedness changed into warnings, threats, and intensions.

I visited Abu Simbel Temples in the Egyptian part of Nubia, a world heritage site on the southern end the River Nile which is said to have been constructed by Ramesses II in the 13th century BC to recognise the success of the Battle of Kadesh for himself and his wife Nefertari. These 2 complex temples were moved in 1968 to enable the flooding of the original plains on which they sat for the Aswan Dam Reservoir. Originally set at the mouth of the Nile it was a warning for non-pharaoh worshippers to proceed no further. Each temple had their own priest and were where wise men who passed on knowledge to the next generation. They were educated in mathematics, politics, reading, writing, but also geometry, astronomy, space measurement and time calculations. The other term for them was *"keeper of key"* but not for Hogwarts.

Georgia Guidestones

Erected in 1980 and filled with numbers, it is a monument the majority know nothing about. Set on the north eastern part of the U.S. state of Georgia and is often referred to as America's Stonehenge but without the £24 adult entrance fee and 2+4=6. Enigmatic perhaps and set on a barren knoll, it is formed from 6 polished granite slabs. Each slab is just over 16 feet tall all conjoined into supporting a 25,000-pound flat stone which adds another 3 feet in height. Set into this capstone is a sun dial marking noontime throughout the year, another channel cut through the stone indicates a celestial pole and another is a horizontal slot which indicates annual travel of the Sun. The principle of Meridiana or gnomon.

Commissioned by R. C. Christian, each vertical slab repeats the same message of population instructions in eight languages. These 10 *guides or commandments* focus on human behaviour and regulation for the population on Earth. Each slab face is written in a different language with the ancient languages on the top stone include Babylonian cuneiform, classical Greek, Sanskrit, and ancient Egyptian hieroglyphics. The modern 8 languages on the lower upright slabs are: English, Spanish, Arabic, Russian, Swahili, Hindi, Hebrew and Chinese.

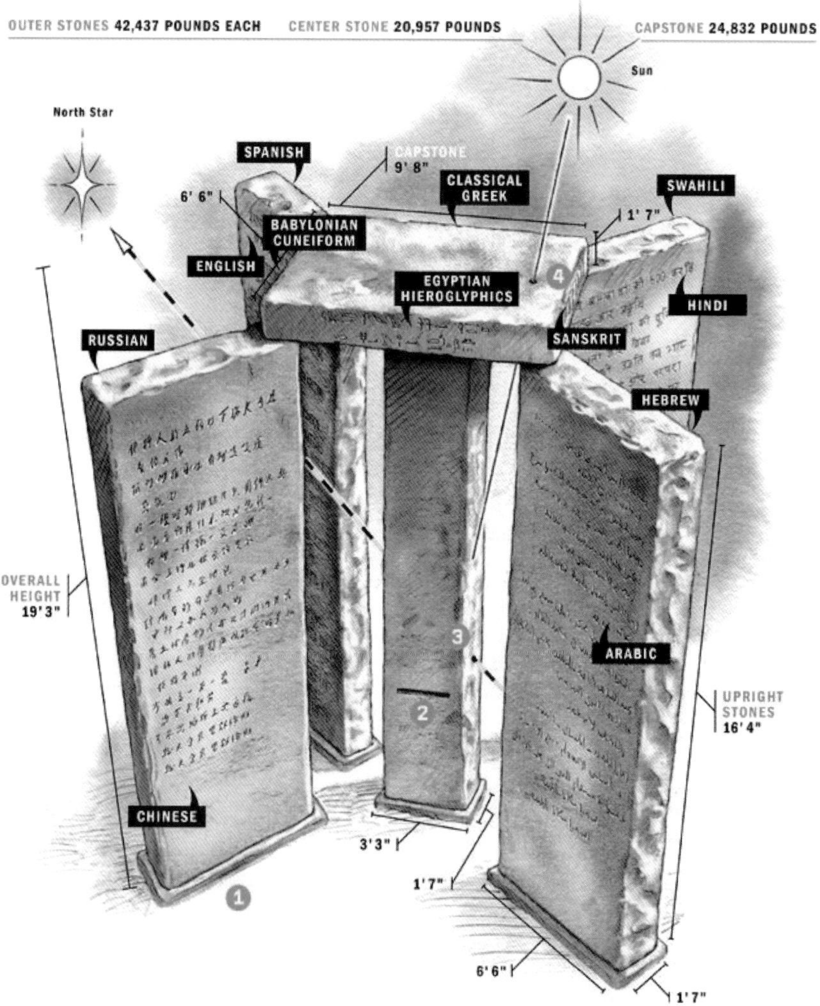

The stones were unveiled on 22nd March 1980. Boris Johnson announced the virus in the UK on the 23rd of March (3) 2020 having decided to do so at a Cobra meeting the day before on the 22nd. Now look at the date it was constructed 22nd March or 223 the inverse of 322.

Various astronomic features exist which include:

- A Circular hole on the celestial pole and north star.
- A horizontal slot indicating the annual travel of the Sun throughout the day.
- A sunbeam through the capstone marking noon throughout the days with the Sun's rays.

The 10 guides or commandments on the 8 faces of the 4 vertical slabs are:

1. Maintain humanity under 500,000,000 in perpetual balance with nature.
2. Guide reproduction wisely — improving fitness and diversity.
3. Unite humanity with a living new language.
4. Rule passion — faith — tradition — and all things with tempered reason.
5. Protect people and nations with fair laws and just courts.
6. Let all nations rule internally resolving external disputes in a world court.
7. Avoid petty laws and useless officials.
8. Balance personal rights with social duties.
9. Prize truth — beauty — love — seeking harmony with the infinite.
10. Be not a cancer on the Earth — Leave room for nature — Leave room for nature.

What do this "guidance" really mean? It is widely understood by government leaders (Mis - Leaders) regarding guidance's that they (1) believe that the world is overpopulated. I would argue its resource distribution is the main issue as some families live in a one-bedroom flat while others have multiple homes, bedrooms never used or personal golf courses. Point (2) on fitness and diversity it is a personal responsibility for your own health. Point (3) is the Jesuit pursuit of Esperanto (translate - Hope) to be a world language. Point (4) attacks faith and tradition. (5) Maritime Admiralty over the Law of the Land. (6) is a New World Order dispute resolution. (7) non-political issues will not be visited. (8) is avoidance of law but notice not its removal. (9) and (10) appears to be just lip service supporting the previous 8 guidance.

The R. C. Christian reference to its creator and commissioner of the stones has led me to some interesting books in my investigations. Those who hate Christianity may jest by using the name, so this can be phonically read as Arse-y Christian or one who has turned away from God. Go back to the year 1614 and 1615, 2 books were published in Germany:

- FAMA FRA-TERNITATIS (The Fame of the Brotherhood of RC)
- Confession Fraternitatis (The Confession of the Brotherhood of RC)

While their release coincides with the death of Shakespeare, these books were heresy and were books used by those who entered the secret society of the Rosicrucian Order or the Rosy Cross (R.C.). The founder of the Rosicrucian's was "Frater C.R.C." who died at the ripe old age of 106.

May 1st is the 121st day of the year (122nd in leap years) in the Gregorian calendar. There are 244 days remaining until the end of the year.

The stones themselves are 19 feet 3 inches which is 221 inches less 10 abutting topsoil and R.C. Christian (EO) - 18+3+3+8+8+9+19+20+9+1+14 = 112. Now add them:
221
112 +

333

The Bavarian Illuminati was agreed on March 22 or 3-22 in 1776, then stablished on the 1st May 1776. The Satanic numbers are in use throughout this structures design.

The Bavarian Illuminati (EO) - 20+8+5+2+1+22+1+18+9+1+14+9+12+12+21+13+9+14+1+20+9 = 221.

Even if you take the 3 words and add them up you get the following in Gematria *Three Twenty-Two* (EO) - 20+8+18+5+5+20+23+5+14+20+25+20+23+15 = 221

- RC (RO) - 9+24 = 33
- Masonry (FR) = 33
- Secrecy (FR) = 33
- Order (FR) = 33

The population reduction is linked to Agenda 21 and the population of Earth 2+1=3.

1776 The Illuminati was set up
1976 The Granite Association of Alberta, the people who carved stone was established

Thomas Paine wrote a pamphlet intitled *Common Sense* and later another work called *The Age of Reason* which was one of the phrases used often by R.C. Christian. If you consider the plaque where the Georgia Guidestones are installed, 1 of them states "LET THESE BE GUIDESTONES TO AN AGE OF REASON". Paine's understanding pertaining to an Age of Reason is the ignoring of all Christian values, a world of non-religion. My research has shown R.C. Christian was Dr Herbert H. Kersten (1920-2005) of 730 Wraywood Drive, Fort Dodge, Iowa 50501. Interestingly the name Kersten is of Dutch origin meaning Christian. The Guidestones were opened on 22nd March 1980 and the 1st of the 3 great American Eclipses was on 21st August 2017. If you calculate the dates and days between the two it amounts to 13,666 days.

Clocks Watches and body parts
Not obvious at first glance but think about the words that are used in anatomy and watches. Hands, movement, dial ring, bezel, band, lug, case, crown, pins, buckle, pusher, dial, sweeping hand, hands, face, back, buckle tongue, cock, balance, screw, pallet, centre hole, hair spring, balance cap, bell(y) button, lock head, signature, oscillating weight, exhibition back, foot hole, date finger, stud. Formal/informal in its dressing, vintage or modern, you can wind it up or allow it to unwind. Timepieces in relation to show have other occultic meanings with the human body, activity, and dress.

AM is a Latin acronym for ante-meridiem which translates to before midday. *PM* is Latin for post-meridiem or after midday. Clocks have and are based on the numerical 12 which is centred upon passing time. Planetary movements are based upon the stars and the stars in turn on constellations and in history have become the 12 signs of the zodiac. Both are wheels of time. A clockface is 2-dimensional but the star system is 3 dimensions. Have you ever thought of why a clock runs in a clockwise direction? Upon a sundial, the sign for midday is at the bottom, because the pointer casts its shadow downwards. The morning hours are indicated on the left and the afternoon hours on the right. In Hebrew, however, they read from right to left and that is the reason for the clockwise reading.

1 hour on a clock demonstrates 1 sign of the zodiac. 12 demonstrates 1 year and 12 zodiac signs which work midmonth as follows:

1	Capricorn	1st hour	January
2	Aquarius	2nd hour	February
3	Pisces	3rd hour	March
4	Aries	4th hour	April
5	Taurus	5th hour	May
6	Gemini	6th hour	June
7	Cancer	7th hour	July
8	Leo	8th hour	August
9	Virgo	9th hour	September
10	Libra	10th hour	October
11	Scorpio	11th hour	November
12	Sagittarius	12th hour	December

Remember that there are 12 inches in a foot; the universe, time, and space are linked. With the 12 being intervals, so is time in relation to days as follows:

1	30 days	1 month	30 cm	1 foot		
2	61 days	2 months	61 cm	2 feet		
3	91 days	3 months	91 cm	3 feet	1 yard	Third of the year
4	121 days	4 months	121 cm	4 feet		
5	152 days	5 months	152 cm	5 feet		
6	182 days	6 months	182 cm	6 feet	1 fathom	Half-year
7	213 days	7 months	213 cm	7 feet		
8	243 days	8 months	243 cm	8 feet		
9	273 days	9 months	273 cm	9 feet	three-quarter year	
10	304 days	10 months	304 cm	10 feet		
11	334 days	11 months	334 cm	11 feet		
12	365 days	12 months	365 cm	12 feet	1 year	

Let us now discuss where the number 12 started. The Sun when viewed from the Earth it is at an angler diameter of 0.5°. Now consider 360° and divide that by 12, the number we are left with is 30 but as there are 24 hours in the day therefore 30 needs to be doubled to 60. If we multiply 60 x 0.5°, we are left with 30 again. Then if we measure the distance of the Sun's diameter in a full year as seen from Earth it theoretically can place 60 Sun's side-by-side to make one sign of the zodiac ,all the spaces between 12 o'clock and 1 o'clock. The Sun's angle to the Earth fits in to the face of a clock 180 times every 3 hours.

But this gets deeper with time itself:

- 60 seconds = 1 minute
- 60 minutes = 1 hour
- 1 hour = 60 Seconds x 60 minutes = 3,600 seconds
- 3,600 seconds x 12 hours = 43,200 the total number of seconds in one half day
- 60 minutes x 24 hours = 1,440 minutes in 1 day

If you remember the number for rotations of the Sun in 1 year, calculated elsewhere in this book, it equates to 322 to 324 revolutions.

Revolutions to Revelations

Revelation chapter 14:1 "*Then I looked and there before me was the lamb, standing on Mount Zion, and with him 144,000 has his name and his father's name written on their foreheads*".

Revelation chapter 14:3 "*And they sang a new song for the throne and before the full living creatures and the elders. No one could learn the song except the 144,000 who had been redeemed from the Earth*". In numeric code before living creatures is a reference to The Four Seasons, and the 1440 minutes within one day, this verse refers to time.

Look closer at the 3,600 seconds x 12 hours = 43,200 (the total number of seconds in one half day). 24 hours or one complete day would bring this number to 43,200 x 2 = 864,000 seconds. Now consider the diameter of the Sun is 864'938. Time has a direct correlation to the size of the Sun.

Things really start to get going when you consider it takes light 1322 seconds to travel the Earth's orbital diameter which is 186,000,000 miles wide and with the speed of light of 186,000 miles per second in my view this is a very scary coincidence given the exceptionally enormous number.

Today's time standard is governed by the Coordinated Universal Time (UTC). The universal time was created in October 1884 at the International Meridian Conference which was held in Washington DC (America), 26 participant countries but only 25 attended as Denmark was absent. The 41 delegates discussed a choice of meridian as the international standard for zero degrees longitude. At the time Greenwich Mean Time (GMT) was chosen as the Prime Meridian was the transit circle at the Royal Observatory in Greenwich, a small village originally just outside London. Adopted 1967 the Coordinated Universal Time (UTC) was the successor to GMT as the world's standard. Until 1911 the French clung to the Paris meridian as a rival to GMT for timekeeping purposes.

If you look at the longitude of the Paris Meridian which is 2° 20 minutes East of GMT and then remember that 2.2 lb = 1 kg. Now look at the Great Pyramid of Giza it has a longitude of 31° 8 minutes East of GMT, if you calculate the difference between the two you are left with 28° 54 minutes and the Moon in terms of time takes 28.53 days for full rotation. Those who governed time on Earth and built the pyramids despite the generations between the two in my view knew exactly what they were doing.

The moon

Let us bring the Moon into the equation of time. The word 'moon' in origin means month and now defined by either 28 or 30 days.

Now that we have the numbers , look at the phases of the Moon which has a total of 8, they appear themselves on a clockface as follows:

	Moon Phases	Time on a clock face
1	New Moon	1.30
2	Waxing Crescent	3.00
3	First Quarter	4.30
4	Waxing Gibbous	6.00
5	Full Moon	7.30
6	Waning Gibbous	9.00
7	Last Quarter	10.30
8	Waning Crescent	12.00

Each phase is named after the position in the full 29.5 days cycle. Many nicknames of the Moon drive from Native American culture which have found their way into methods of timekeeping annually and work with a Gregorian calendar.

January or 'Wolf Moon' was named after the howling of hungry wolves lamenting the scarcity of food at the height of winter, another term for this time of year is 'Ice Moon'. February or 'Snow Moon' had its name from the snowy wintry weather of North America, 'Hunger Moon' is another common name for this period. March or 'Worm Moon' is for when the soil starts to thaw in 6-week trails. April or 'Pink Moon' from the native Americans, it is at home for the early blooming of wildflowers, while 'Flower Moon' in May is the start of the blooming spring. June known as the 'Strawberry Moon' and in Europe known as the 'Rose Moon' due to the on setting of the summer heat. July is the 'Buck Moon' when a male deer shed their antlers after the rutting season; other terms are 'Hay Moon' and 'Thunder Moon' due to increased electrical storm activity. 'Sturgeon Moon' in August is when North American tribes had fish in great abundance, other terms include 'Red Moon' due to summer heat haze. The Moon in September is referred to as the 'Full Corn Moon' due to the harvesting of crops hence the term 'Harvest Moon'. October is the 'Hunters Moon' as it is the main month to hunt summer fattened land animals, due to the fields being bare they could be seen which would be easier for hunters. The origin of 'Beaver Moon' in November according to the Old Farmer's Almanac, it states that it comes from folklore and it is when beavers take shelter in their lodges to prepare for winter. December's 'Cold Moon' derives from the coming winter and the lengthening of winter hours. 'Blue Moon' is a term for a second moon in any 1 calendar month or the 13th moon in the last month of the year or the 3rd full moon in any 1 season.

Again, I ask you to put this book down and find yourself a watch that simple thing which most households have in one form or another...off you go. Space, time, and direction will all be shown to you now as they have a direct relationship. If you are ever lost and the Sun is shining, lay the watch horizontally and align the hour hand of the watch with the direction of the Sun. The middle point between the alignment of the Sun with the hour hand, and the 12 o'clock position on the dial, indicates South. Positioning the rotating bezel so that it points South, will then allow you to read other compass directions. This is why expensive watches are called navigators.

DVDs and Time

Compact discs or DVDs (Digital Versatile Disc) are high-resolution audio-visual recording devices first released in 1996. This 3-dimension medium stores digital data, the diameter of which is 12 cm (4.7 inches) with a thickness of 1.2 mm (0.047 inches), they weigh 16 g (0.56 ounces). They record different speeds for example SLP = Super Long Play, EP = Extended Play, LP = Long Play, SP = Standard Quality Mode and HQ = High Quality Mode. Now what if I was to tell you that these recording devices like watches or codex methods are entrenched with measurement, months, and time. Let us explore:

Speed	**Time (minutes)**	**Day**	**Metric (cm)**	**Imperial (Feet)**	**Zodiac (Month)**
HQ	60	60th	60	2	2nd
SP	121	121st	121	4	4th
LP	182	182nd	182	6	6th
EP	243	243rd	243	8	8th
NA	304	304th	304	10	10th
SLP	365	365th	365	12	12th

CERN and Marriage

This scientific Institute is investigating the moment of creation but look at the word *moment,* which is a medieval unit of time. CERN is derived from the acronym for the French "**Conseil Européen pour la Recherche Nucleaire"**, or European Council for Nuclear Research. This SUBATOMIC PARTICLE ACCELERATOR lays on coordinates of 46° 14′ 6″ North and 6° 2′ 42″ East of GMT (666 appears). If you factor in its first operational date of the 10th September 2008 another problem appears. For those females who have proposed to their unsuspected boyfriends will know about leap years. The date is important as it is the day following 254th day of the leap year and is synonymous for this day is now known as 9-11. Just another number coincidence.

The Sun runs all biological systems in our universe including our DNA but for now ponder these numbers, they relate to time, space, and measurement. The question is could our DNA evolve as the universe expands? Would it therefore have a direct relationship with human evolution?

10:10

I am asking you again to put this book down and do some research. Simply type into Google (other and far better search engines are available) "watch adverts", look carefully at the times those watches display. Stop reading, do as I ask...otherwise this paragraph will make no sense in your realisation. Those of you who are truly engaging in the conversation may have noticed my indication of the time. So why 10:10? Time is a linear existence for those of us alive and a watch dictates what time you must be at a certain place. Ever heard the term "1010 till I see you again"? This is about meeting on time and your life ticks away. The esoteric meaning represents the two arms of Jesus on the cross and at the foot of Jesus with the sign INRI. This has subsequently been replaced with the logo of the watch manufacturing company to draw the eye from the two arms to the foot of the watch face.

All said, you will never look at time the same again.

Chapter 8 - Shipping, Serial Killers and Adrenochrome

In 2014 I came across the word *Adrenochrome*, it was described as the elixir of life. The research, however, took me into the dark recesses of human behaviour which culminated in mentioning it in a live talk I gave (within the small awake community) to The Alternative View Network in 2020. People reacted to what I was saying and that lead into Luciferase, ID2020 and other awful elements that I discovered prior to the vaccine rollout. It was the biggest personal contact from people I had received from any lecture I gave up to that time.

If this is new to you, I can guarantee it will become a word that will horrify you and future generations! I will take you through this minefield carefully, but the occult facts I found are unnerving. I can handle the truth, which sounds like a strange thing to say but as a researcher you must be willing to "go with it" (chant of 3), to explore things not knowing its destination.

I could have placed these findings into the chapter on blood, but instead I have added this element on the subject here. There are many tales in history of vampirism or the bathing in and consuming blood for youthfulness. Folklore tales from Eastern Europe predominantly from Germany are steeped in myth and legend, these stories, however, go much further back in history. They are found in Greek mythology and the god Apollo. The myth goes something like this: *In a fit of rage Apollo the creator beguiled a human called Ambrogio an Italian born adventurer. Apollo cursed him and decreed that his skin would burn if he ever was touched by sunlight, he then through a series of unfortunate events gambled his soul away to Hades the god of the underworld. Apollos's sister Artemis the goddess of the Moon also cursed him should he touch silver. Over time she regretted her curse but made things worse by giving him immortality. She was also the goddess of hunting by moonlight so gave him such speed and knowledge of stalking he was only second to herself.* This myth has all the elements that went into Bram Stoker's tale, *Dracula*. There are cults that still believe that Artemis still resides somewhere in Florence (Firenze).

The Chinese also had a mythical belief in a similar entity called *Jiangshi* but the interesting fact here lays in *salt* or *sal*. They believe that if such a creature is pursuing you then you should throw salt behind you so that the assailant would be obliged to stop and count every grain, allowing you to flee. There is also the fact that salt is used in the West to keep these spirits from crossing the path or doorway or to also stand within a salted circle with demonology. The salt concept is now a planetary belief. Salt over your shoulder while cooking should you spill any is for protection and is still done to this day.

Human blood has salt as a component, but it also has stress hormones under specific circumstances with the flight or fight hormone called *adrenaline*. Released by the pituitary gland within the brain in high stress or painful events, this hormone creates enormous strength, heightened awareness, and instant physical changes; with the brain speeding up reaction time. It is used to treat anaphylaxis, cardiac arrest, and it stops non-arterial bleeding. It is used in an Israeli bandage (bandages for serious bleed outs, which you can buy on ebay – I keep them in my car). Adrenaline is also addictive which drives the term adrenaline junkie. This was the subject of limited research between 1950s and 1970s as a potential cause or cure of schizophrenia. The current medical application for the compound and related derivative compounds, *carbazochrome* a haemostatic medication.

Adrenaline, however, is a strange hormone and its component chemicals are as below:

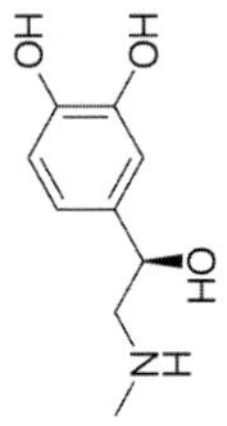

EpiPen's are used in the case of sting reactions or allergies, these pens are filled with adrenaline as it is fast and immediate in its effectiveness and in this case, it is called *Epinephrine* (medication). However, there is a more sinister nature to adrenaline. If oxygen floods the chemical, it alters and adrenalizes the cell walls as it oxidates from red into a deep brown colour, the chemical that results in its formulation is *adrenochrome*.

Here is the chemical compound of Adrenochrome:

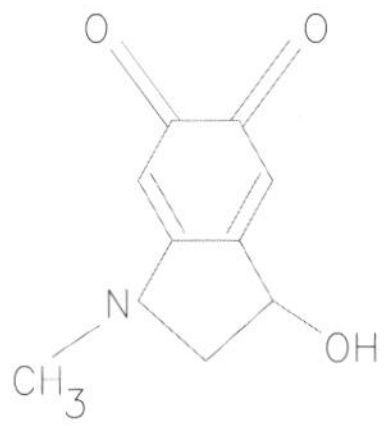

Chemically, adrenochrome and adrenaline are very reactive substances. Some of the changes produced by adrenochrome may persist for several days, and in some cases, the effects lead to disastrous results. Studies are proving this may indeed halt and reverse the appearance of aging while those consuming it can get a euphoric high, it is reputed to be far stronger than THC, LSD, MDMA or other hallucinogenic drugs. Withdrawals, however, create accelerated deterioration to the stage of which you were at before you started taking this drug.

Adrenochrome ("*Walnut Sauce*")

We now delve once more into the occult and the symbol which has leaked into literature and movie making. Movies like the *Matrix* and *Alice in Wonderland* have a common occurrence which happens to be about *"following the white rabbit"*. White rabbit is referred to amongst drug users of LSD and Opium, to see beyond the illusion of this reality. It is a gateway drug that is said to open the door of your reality and lets the true reality become visible and interact with it. Drugs that have the means to willingly look beyond the obvious, and experience the hidden, or not easily visible. In life's normal experience metaphorically, it came from the gnostic teachings of the church and were in reference to man's spiritual pursuit of the meaning of life and his futility in trying to find it. A tattered remnant of this story was found in the excavations in ancient Egypt. The story revealed that in the pursuit of the "white rabbit", at some point the rabbit stops in its tracks and turns around to attack you with its long claws and sharp teeth. Remember the *Monty Python* sketch regarding the vicious white rabbit, you will see the connection. Native Americans believed that a white rabbit was a symbol of a spiritual guide to enlightened energy.

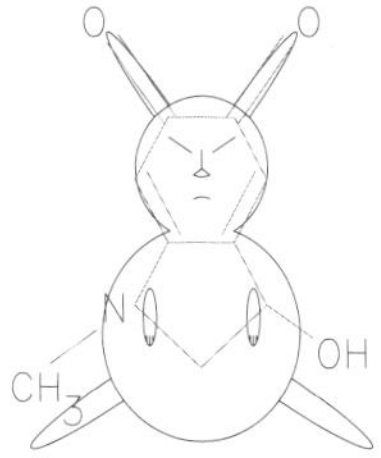

Here is the esoteric combination between the white rabbit and Adrenochrome:

The more youthful the victim of adrenaline harvesting is, the stronger this effect is said to be. It is also believed that there is predation of this with kidnapping and trafficking children for harvesting of this compound. The occult code name is 'walnut sauce'.

The Book of Leviticus in the Bible, Leviticus 17:10-14 to be precise mentions the act of ingesting blood (and thus the adrenochrome borne by that blood) was prohibited scripturally: *"whatsoever man there be of the house of Israel, or of the strangers that sojourn among them, that eats any manner of blood, I will set My face against that soul that eats blood, and will cut him off from among his people. For the life of the flesh is in the blood; and I have given it to you upon the altar to make atonement for your souls; for it is the blood that makes atonement by reason of the life. Therefore I said unto the children of Israel: No soul of you shall eat blood, neither shall any stranger that sojourns among you eat blood. And whatsoever man there be of the children of Israel, or of the strangers that sojourn among them, that takes in hunting any beast or fowl that may be eaten, he shall pour out the blood thereof, and cover it with dust. For as to the life of all flesh, the blood thereof is all one with the life thereof; therefore, I said unto the children of Israel: Ye shall eat the blood of no manner of flesh; for the life of all flesh is the blood thereof; whosoever eats it shall be cut off*". This is the basis for the halal meat which is drained of blood.

Jeffrey Epstein's arrest has exposed the fact that paedophile rings really do exist and that the problem is larger than most people had ever thought to be the case. People are beginning to learn that Ritualistic Satanic Abuse (RSA) exists with children being used for various ceremonies, sacrifices, and rituals. Children are hurt and even killed, and many who carry out the heinous acts involved seem to be from the upper echelons of political, corporate, and cultural power globally.

The chemical reaction that produces adrenochrome is described in a whitepaper written at California State University, Northridge, entitled "Essay for Superoxide Dismutase Activity Using the Enzyme Inhibition of the Oxidation of Epinephrine", the document follows that O2 builds in the solution, the formation of adrenochrome accelerates because O2 also reacts with epinephrine to form adrenochrome. In the final stages of the reaction, when the epinephrine is consumed, the adrenochrome formation slows down and stops.

As well as a high rejuvenation, the desired outcome for those who ingest this adrenochrome-laden blood allows the body to conduct cell mitosis closer to the level it does for minors, improves regeneration of muscle tissue, and prevents chromosomes from being pulled apart during mitosis (cancer) which strengthens them. The more youthful the victim of adrenaline harvesting is, the stronger this effect is.

In 1954 an essay by Aldous Huxley *The Doors of Perception* discusses the possibility of Adrenochrome, which in this work is called mescaline. In the movie *A Clockwork Orange* (a 1962 novel by Anthony Burgess), the movie industry tells us all about this with the Moloko Plus drink, a mixture of Adrenochrome and milk. Adrenochrome is also mentioned in *Fear and Loathing in Las Vegas* (FALILV) a 1971 book by Hunter S Thompson and in the BBC Crime Drama called *Lewis* (2007 2nd episode called "Whom the Gods Would Destroy").

An interesting company called Ambrosia was set up by Jesse Karmazin, they controversially were charging $8000 to fill an adult's veins with young blood. They operated out of 5 cities in the US claiming it gave people superhuman powers. In Greek mythology *Ambrosia* translates to immortality.

Adrenochrome in simple gematria is 1+4+18+5+14+15+3+8+18+15+13+5 = 119 or reversed 911.

More gematria with Mr Thompsons work for you:

- Fear and Loathing (FR) - 6+5+1+9+1+5+4+3+6+1+2+8+9+5+7 = 72
- Hunter S Thompson (FO) = 72
- Hunter (EO) = 86
- Loathing (EO) = 86
- Las Vegas (EO) = 86

Interestingly the movie FALILV was released exactly 1350 weeks after the book was published but then you take "fear and loathing" in (EO) - 6+5+1+18+1+14+4+12+15+1+20+8+9+14+7 = 135. Look up the graphics card named Qualcomm *Adreno* 530 and then Google's browser called *Chrome*. Combined they make the word Adreno-Chrome.

Cartoons

Even *Monsters, Inc.* got in on the act. Consider the structure of the storyline; it involves adult monsters terrifying children to the point where the energy they create through fear is harvested, that energy is used so the monsters can exist.

Green

HR3C is the call sign name given to the Ever Given (EG) container ship stranded in the Suez Canal from 23rd March to 29th March 2021. This vessel is owned and operated by the Evergreen Company. Hilariously the code (Evergreen) given to the vessel is also the Secret Service code given to Hillary Diane Rodham Clinton. The coincidences do not just stop there, the first vessel to go to the assistance of the Ever Given was the BARAKA1. The following 3 were from the same shipping company and had the following names MOSAED 1, 2 and 3. The vessel EG which was only 3 years old is registered in Taiwan, operated by Walmart, and wait for it, the Clinton Global Foundation. Ghislaine Maxwell's husband Scott Borgerson is the CEO of CargoMetrics a data-analytics company for maritime trade and shipping. Those who follow the esoteric will see this obvious connection.

Evergreen Life

This is the name given to the National Health Service (NHS) in the United Kingdom for the storage of patient medical records application (to be used on your phone or computer). It is supposed to enable you to actively manage your health and fitness and well-being, a tracking system of your lifestyle goals which includes booking appointments with your general practitioner, ordering repeat medication and downloading your medical records.

4 types of Serial Killers and 4 reasons for being not guilty

We lead into the law for a moment. There are 4 reasons that have been successful in someone not being found guilty in court for murder:

1. Menstruation
2. Having a brain tumour – *amygdala* – area in the brain affecting behaviour
3. Taking anabolic steroids (testosterone derivatives)
4. Eating junk food; Dan White used it successfully and called it the *"Twinkie Defence"*

The commonality here is the hormone imbalance which has been demonstrated for altering an individual's character.

In this book I look at movies and try to find the occult within them, during my research I found an interface with *Leatherface* in *The Texas chainsaw Massacre*. Though the character in the film was under the zodiac sign Aries, it did lead me to other characters such as *Freddy Krueger* (*A Nightmare on Elm Street*) a Sagittarius, *Samara Morgan* (*The Ring*) a Pisces, *Jack Torrance* (*The shining*) a Gemini and *Pinhead* (*Hellraiser*) a Virgo. The outcome of my research was the realisation that the movie industry depicts serial killers to a larger extent as Sagittarius, Gemini, Pisces, and Virgo.

The reasoning seemed odd until I painstakingly reviewed real serial killers and looked at the signs of the zodiac under which they were born:

Ted Bundy	Sagittarius
Thierry Paulin	Sagittarius
George Chapman	Sagittarius
Ahmad Suradji	Sagittarius
Jake Bird	Sagittarius
Timothy Krajcir	Sagittarius
Yvan Keller	Sagittarius
Edmund Kemper	Sagittarius
Dennis Nilsen	Sagittarius
Rose West	Sagittarius
Derrick Bird	Sagittarius

Carlton Gary	Sagittarius
Jeffrey Dahmer	Gemini
András Pándy	Gemini
Daniel Lee Siebert	Gemini
Francisco García Escalero	Gemini
David Berkowitz	Gemini
Richard Chase	Gemini
Samuel Little	Gemini
Danny Rolling	Gemini
Arthur Shawcross	Gemini
Kenneth Bianchi	Gemini
Wayne Williams	Gemini
Peter Kürten	Gemini
Ted Kaczynski	Gemini
Lorenzo Gilyard	Gemini
Mary Bell	Gemini
Peter Sutcliffe	Gemini
John Norman Collins	Gemini
John Wayne Gacy	Pisces
Mikhail Popkov	Pisces
Charles Cullen	Pisces
Donald Henry Gaskins	Pisces
Randy Stephen Kraft	Pisces
Alexander Spesivtsev	Pisces
Richard Ramirez	Pisces
Denis Pischikov	Pisces
Michel Peiry	Pisces
Lesley Bailey	Pisces
Ottis Toole	Pisces
Aileen Wuornos	Pisces
Dennis Rader	Pisces
Christopher Wilder	Pisces
Gary Ridgway	Pisces
Doug Clark	Pisces
Robert Napper	Pisces
Peter Manuel	Pisces
Vladimir Kuzmin	Virgo
Edward Gein	Virgo
Richard Biegenwald	Virgo
Terry Blair	Virgo
Dean Carter	Virgo
Adnan Çolak-	Virgo
John Justin Bunting	Virgo
Andrew Cunanan	Virgo
Albert DeSalvo	Virgo
Henry Lee Lucas	Virgo
Paul Bernardo	Virgo

Gerald Stano	Virgo
Marybeth Tinning	Virgo
Harrison Graham	Virgo
Richard Angelo	Virgo
Rodney Alcala	Virgo

CBAN 348 9

The popular fictional showtime TV series called *Dexter* has an uncanny connection to Evergreen. Dexter a serial killer, hunts down criminals and other serial killers and dispatches them. In episode 12 of Series 1, Dexter has a child hood memory of being rescued by a police officer from a shipping container number CBAN 348 9, together with another child with the container's floor being deep in blood. The allegory is that he was a child who suffered trafficking and was rescued from the Port of Los Angeles in Miami. In the book there is no specific mention of the container that the child Dexter was rescued from but in the TV series the logo on the side of the shipping unit had the word "Evergreen". The later books reference the cult of *Moloch* and human sacrifice. The inference is that Dexter was abused and became an abuser but in this clip no producer would ever involve a corporate logo in this way for fear of being sued.

While Dexter was investigating, he was staying in the Marina View Hotel and he spilt blood inside room 103. He tuned the radio station to 103 FM, bookmarked the hotel's Bible to Leviticus 10:3 and later he set the clock at Santa's Cottage to 1:03 pm. Dexter as a child was found on October 3rd or 103 and 103 is the next prime number after 101. I will leave Leviticus for you to investigate yourselves, have fun.

Interestingly fact - Dexter was a Virgo.

More Number Coincidences

JFK was given a white rabbit by a magician in 1961. Abraham Lincoln was sent two white rabbits as a gift in 1862. Both were Freemasons and both understood the deep state. I leave the question that these gifts of white rabbits imply their knowledge of the rabbit holes. Both were assassinated and interestingly the numbers regarding their deaths had startling similarities which are as follows:

Lincoln = 110	**Kennedy = 111**
Born 1809 - 1+8+9 = 18 and 1+8 = 9	Born 1917 - 1+9+1+7 = 18 and 1+1 = 9
Elected to Congress in 1846	Elected to Congress in 1947
Elected for presidency 1860	Elected for presidency 1960

- Both had a seven-letter name
- Both had a 5-syllable name
- Both presidents concentrated on Civil Rights
- Both wives lost children while in the White House
- Lincoln's son / William Wallace Lincoln-7 letters
- Kennedy's son /Patrick Bouvier Kennedy-7 letters

- Lincoln's secretary was called Ms Kennedy and Kennedy's secretary was called Ms Lincoln
- Both presidents were shot on a Friday
- Both presidents were shot in the head
- Both were shot in front of their wives
- Southerners assassinated both
- Southerners succeeded both
- Andrew Johnson born 1808 succeeded Lincoln
- Lyndon B Johnson born 1908 succeeded Kennedy
- John Wilkes Booth shot Lincoln was born in 1838
- Lee Harvey Oswald who shot Kennedy was born in 1939
- Both assassins were known by three names
- Both assassins' names are comprised of 15 letters
- Lincoln was shot in the Ford's Theatre and Kennedy was shot in a vehicle called a Lincoln Continental made by Ford Motor Company
- Booth and Oswald were both assassinated before they could go to trial

At the end of this chapter, I leave you to think about the fact that numerically if a jury of 12 decides to execute and kill a serial killer, Earth's killers increase!

Chapter 9 - Magic Squares, Turtles and Demons

While reading this chapter you will need a calculator, so find one before reading these pages. This chapter gets a little complicated unless your mental arithmetic is accurate.

Part of my professional career was to deal with the substantial fitouts of office buildings which, on occasion were being occupied by Chinese companies who extensively used a Feng Shui Master. Pronounced *"feng shway"*, it is an ancient traditional practice which claims to use energy forces to harmonise human physical surroundings, environment, and body. The energy is called *Qi*.

The Chinese believe these cosmic currents (the circulating life force whose existence and properties are the basis of most Chinese philosophy and medicine). The beliefs in this culture are so strong, the very thought of existing without this practice is quite rare and even considered reckless. In the West it is considered pseudoscientific but over the decades as I dealt with these Feng Shui Masters, I was able to put together some of the missing pieces which, I now share with you.

The *Lo Shu* (Chinese- meaning - *River Writing or Map*) is named after a Chinese river populated by turtles. It is also called the *Divine Turtle* in esoteric schools as the patterns of the shell is segmented. There is an ancient Chinese legend about it that goes something like this:

"About 3000 years BC a great flood washed through an area near a river called the Lo Shu. The river was used for sacrificing human children, animals, fish, and reptiles to the gods for the benefit of those carrying out this barbaric practice. One of the sacrifices offered was a big turtle which emerged from the river itself. At the point before sacrifice, it was noted that on the back of the Testudines shell (tortoise – turtle) was a 3 x 3 grid with dots in each section that seemed organised. Deemed to be a messenger from the gods these number patterns were explored and so was born the legend of the magic square (LO-SHU)".

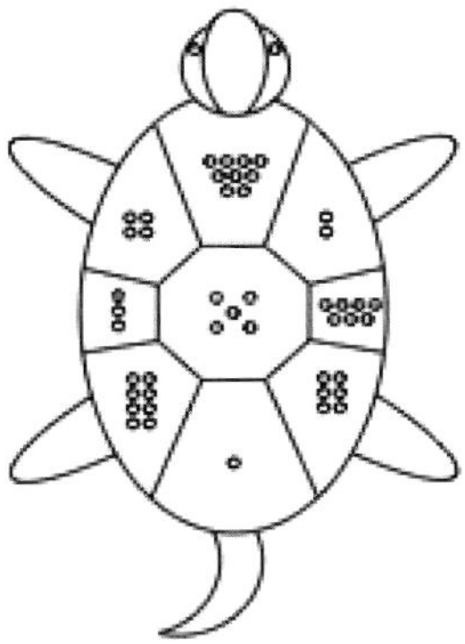

The turtle shell pattern was said to have demonstrated occult numbers and was extrapolated into 9 boxes and each box represented the then nine provinces of China. The 9 formations on the shell revealed itself into a layout as follows:

4	9	2
3	5	7
8	1	6

This symbolic wisdom from the gods is said to have born Yin and Yang, the flow of Qi, the Chinese Principles of Reality, Taoist Cosmology, Feng Shui, the I Ching, and traditional Chinese medicine. We will continue with this Lo Shu later, but first let us look at the mathematics of the squares.

Magic squares have numbers in cells, and these cells therefore have energy with direct frequency.

These are now known as *planetary squares*, there are 18 variants of the square above using the 9 numbers. The magic square is divided into the 9 squares which contain the number from the turtle's shell pattern. The figure inside each box when added to others vertically, horizontally, and diagonally strangely add up to the same number. The following magic square is a 3x3 index (boxes with the calculation numbers) adding to 15 in any given direction. Look back at the turtle shell and do the mathematics.

4	9	2	= 15
3	5	7	= 15
8	1	6	= 15
= 15	= 15	= 15	

The corner numbers are even (2,4,6,8) and the odd numbers present themselves as a sign of the cross in the centre. You can create these squares with vast numbers. If used within a small grid, you see where *Sudoku* (Japanese translation – *The Numbers* or *Digits Must Remain Single*) comes from. For the maths hounds among you, it works on the basic equation of:

$$SUM = \frac{M(M^2 + 1)}{2}$$

Now that you understand what a magic square is and its origin, here is a magic square I calculated with the total sum of 123 using a 6x6 format. To be able to do this is important and as we work through this chapter you will see why.

31	9	8	22	27	26
11	34	3	29	25	21
33	5	10	24	23	28
4	36	35	13	18	17
38	7	30	20	16	12
6	32	37	15	14	19

The Steps of Yu

Returning to exploring the Lo Shu now that you understand the mathematical principles. The *Steps of Yu* is sometimes referred to as *Yubu* (Chinese meaning *pace, paces*, or *steps*) which is a Taoist/Daoist belief system on movement of humans based on numbers and magic squares of the Lo Shu turtle. Legend has it that Yu was a legendary founder of the Xia dynasty (2070 BC) who worked so long and hard fighting mythical floods that he became partially paralysed.

To this day the Taoist/Daoist priests have a mystic path they trace in places of worship called the Yubu which is to *"step forward"* but involves the dragging of the back foot the representation of paralysis. The direction of travel across these places of worship follows the number patterns of the Lo Shu, and the 9 provinces of ancient China.

The Yubu or direction of travel starts with the number 1 and traces over to the number 2, then 3.

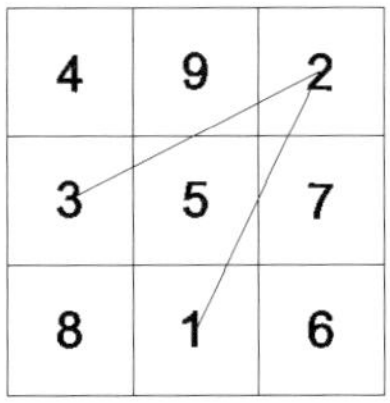

This continues until all the numbers are reached thereby returning to number 1.

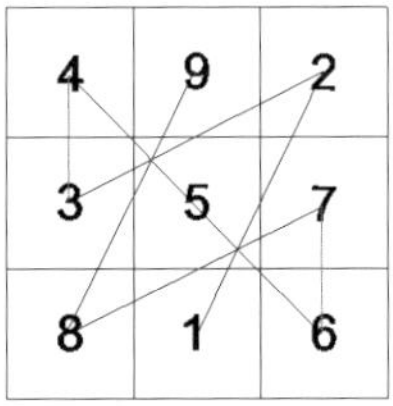

Taking the number sets linear in the boxes of the Lo Shu we get 492 top, 357 middle and 816 bottom; these number sets have esoteric meanings in Chinese religion and magic. The method found its way into the *Grimoire* (French *Book of Spells*) work in the West. The pattern that remains (sometimes called *"linear algebra"*) is then used in magic; we will review this later.

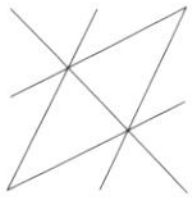

These were unknown to Latin Europe and did not get traction until the 11th century in *"Kitāb tadbīrāt al-kawākib"* (*Book on the Influences of the Planets*) written by Ibn Zarkali of Toledo Al-Andalus, as planetary squares.

The *JAINA Square* was found in the Khajuraho Temple Northern India, believed to date from the 11th century CE and proves that China was not the only place this concept was found.

7	12	1	14
2	13	8	11
16	3	10	5
9	6	15	4

Known also as the *pandiagonal magic square* or *panmagic square* which is a magic square with the additional property that the broken diagonals (also *diabolic magic square),* due to its intricacy and genius. Firstly, the basics of each line is equal to 34 in every direction. Every 2x2 grouping adds up to 34 and all 4 corner squares add up to 34. In fact, there are 52 ways or patterns that can be made within the square that adds up to 34, it truly is the most perfect magic square.

An artist called Albrecht Durer, lived in Germany circa 1500 and depicted magic squares in his work including sculpting, paintings, and etchings. One of his most famous is on an etching from 1514 called *Melencolia I.*

16	3	2	13
5	10	11	8
9	6	7	12
4	15	14	1

All linear squares add to 34. If you then we do the Lo Shu trace, a pattern emerges.

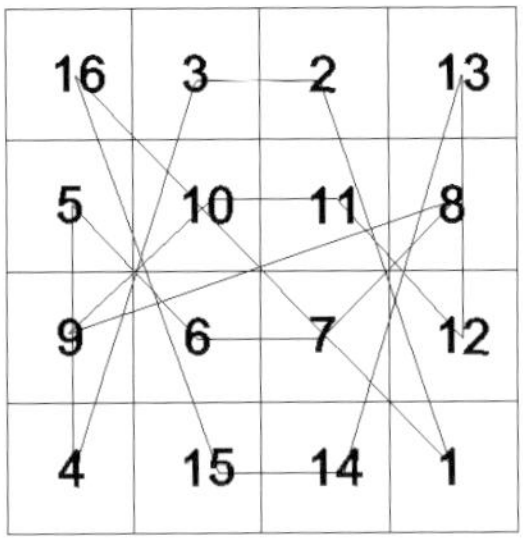

At this point I am going to move into letters for a while but remember the Lo Shu trace as it is important as we move into demonology.

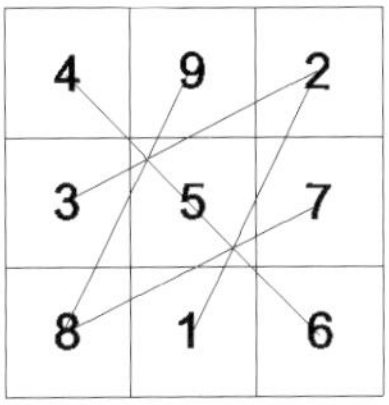

Let us look at the mathematics which sits behind the magic square itself.

4	9	2
3	5	7
8	1	6

We can break the pattern by looking at the numbers 1 through 9 (which makes up the square). Since pyramids exist in letters themselves this is the base patterns we use. By writing the numbers out (1-9) in a diamond (4 pyramid) grid we get the following:

		1		
	4		2	
7		5		3
	8		6	
		9		

Taking the numbers which sit in the periphery boxes such as 1 and 9 and swapping the positions with their counterpart number e.g., 1 swap with the 9, while 3 swaps with 7.

		9		
	4		2	
3		5		7
	8		6	
		1		

If you pull the numbers into the centre, you will see the square appear. By removing the five squares on each side of the periphery you end up with a 3 square grid consisting of 1 to 9 which adds up to 15 in any given direction.

	4	9	2	
	3	5	7	
	8	1	6	

"*Intelligentia Saturni*" created the calculations with Hebrew letters which sum up to 3, 9, 15, 45. These values as before are calculated by writing out the names in Hebrew and then adding up the value of each included letter, as each Hebrew letters can represent both a sound and a numerical value. Names associated with Saturn mentioned previously have numerological values of 3, 9, or 15. The names of the *Intelligence of Saturn* and the *Spirit of Saturn* have a value of 45. The metal associated with Saturn is lead.

- 3 = the boxes on each side
- 9 = the total number of boxes
- 15 = what each direction adds to
- 45 = the sum of the total of all boxes

4	9	2	=15	
3	5	7	=15	=45
8	1	6	=15	

15 and 1+5 = 6 and 666
And 111 = 3 boxes
And 111+111 = 6 for the total of 2 sides
111+111+111+111 = 12 and 1+2 = 3
9 = number of boxes
45 also adds 4+5 = 9
We have 3 and 6 and 9 throughout. These are the Tesla numbers.

Building the Mars Magic Square

				1				
			6		2			
		11		7		3		
	16		12		8		4	
21		17		13		9		5
	22		18		14		10	
		23		19		15		
			24		20			
				25				

Bring inwards toward the centre of the square numbers 1, 5, 21 and 25.

		6		2		
	11		7		3	
16		12	25	8		4
	17	5	13	21	9	
22		18	1	14		10
	23		19		15	
		24		20		

Afterwards swap the following number 6, 2, 4, 10, 20, 24, 22 and 16 with each other.

11	24	7	20	3	=65	
4	12	25	8	16	=65	
17	5	13	21	9	=65	=325
10	18	1	14	22	=65	
23	6	19	2	15	=65	
=65	=65	=65	=65	=65	=325	

Seal of MARS – Geburah - Iron

11	24	7	20	3
4	12	25	8	16
17	5	13	21	9
10	18	1	14	22
23	6	19	2	15

The numbers associated with Mars are 5, 25, 65, and 325. This is because:

- Each row and column of the magic square contains 5 numbers
- The square contains 25 numbers total, ranging from 1 to 25
- Each row, column and diagonal add up to 65
- All the numbers in the square add up to 325. The names of *The Intelligence of Mars* and *The Spirit of Mars* adds to 325. These values as before are calculated by writing out the names in Hebrew and then adding up the value of each included letter, as each Hebrew letters can represent both a sound and a numerical value

Seal of Jupiter – Chesed - Tin

4	14	15	1
9	7	6	12
5	11	10	8
16	2	3	13

The numbers associated with Jupiter are 4, 16, 34, and 136 as follows:

- Each row and column of the magic square contains 4 numbers
- The square contains 16 numbers total, from 1 to 16
- Each row, column and diagonal add up to 34
- All the numbers in the square add up to 136
- Names associated with Jupiter all have numerological values of 4 or 34. The names of *The Intelligence of Jupiter* and *The Spirit of Jupiter* adds to 136. This follows the same explanations as Mars above

Seal of Sun (Sol) - Tiphereth - Gold

6	32	3	34	35	1	=111
7	11	27	28	8	30	=111
19	14	16	15	23	24	=111
18	20	22	21	17	13	=111
25	29	10	9	26	12	=111
36	5	33	4	2	31	=111
=111	=111	=111	=111	=111	=111	=666

Numbers associated with the Sun are 6, 36, 111, and 666. This is because:

- Each row and column of the magic square contains 6 numbers
- The square contains 36 numbers total, from 1 to 36
- Each row, column and diagonal add up to 111
- All the numbers in the square add up to 666
- The Hebrew value is 111

Seal of Venus - Netzach - Copper

22	47	16	41	10	35	4	=175
5	23	48	17	42	11	29	=175
30	6	24	49	18	36	12	=175
13	31	7	25	43	19	37	=175
38	14	32	1	26	44	20	=175
21	39	8	33	2	27	45	=175
46	15	40	9	34	3	28	=175
=175	=175	=175	=175	=175	=175	=175	=1225

Numbers associated with Venus are 7, 49, 175 and 1225. This is because:

- Each row and column of the magic square contains 7 numbers
- The square contains 49 numbers total, from 1 to 49

- Each row, column and diagonal add up to 175
- All the numbers in the square add up to 1225
- The Hebrew value is 1225

Seal of Mercury - God - Quicksilver

8	58	59	5	4	62	63	1	=260
49	15	14	52	53	11	10	56	=260
41	23	22	44	45	19	18	48	=260
32	34	35	29	28	38	39	25	=260
40	26	27	37	36	30	31	33	=260
17	47	46	20	21	43	42	24	=260
9	55	54	12	13	51	50	16	=260
64	2	3	61	60	6	7	57	=260
=260	=260	=260	=260	=260	=260	=260	=260	=2080

Numbers associated with Mercury are 8, 64, 260, and 2080. This is because:

- Each row and column of the magic square contains 8 numbers
- The square contains 64 numbers total, from 1 to 64
- Each row, column and diagonal add up to 260
- All the numbers in the square add up to 2080
- The Hebrew Value is 2080

Seal of Moon (Luna) – Yesod - Silver

37	78	29	70	21	62	13	54	5	=369
6	38	79	30	71	22	63	14	46	=369
47	7	39	80	31	72	23	55	15	=369
16	48	8	40	81	32	64	24	56	=369
57	17	49	9	41	73	33	65	25	=369
26	58	18	50	1	42	74	34	66	=369
67	27	59	10	51	2	43	75	35	=369
36	68	19	60	11	52	3	44	76	=369
77	28	69	20	61	12	53	4	45	=369
=369	=369	=369	=369	=369	=369	=369	=369	=369	=3321

The numbers associated with the Moon are 9, 81, 369, and 3321. This is because:

- Each row and column of the magic square contains 9 numbers
- The square contains 81 numbers total, ranging from 1 to 81
- Each row, column and diagonal add up to 369
- All the numbers in the square add up to 3321
- The Hebrew value is 3321

A *SATOR Square* is also known as a *ROTAS Square* containing five-words Latin palindrome. The earliest forms were carved into stone and were found in Pompei, Rome, France, and England. You can read the squares horizontally beginning in the top left corner, horizontally beginning in the bottom right corner, vertically beginning in the top left corner, and vertically beginning in the bottom right corner. It is a two-dimensional word square with a full directional palindrome (the same word forward and backwards).

S	A	T	O	R
A	R	E	P	O
T	E	N	E	T
O	P	E	R	A
R	O	T	A	S

A good example was found both in Oppède in France and Saint Peter's ad Oratorium in Italy. It consisted of a 5x5 square with 5 letter words totalling 25 letters.

R O T A S	S A T O R
O P E R A	A R E P O
T E N E T	T E N E T
A R E P O	O P E R A
S A T O R	R O T A S

The meaning of the words are as follows:

- **Sator -** (nominative or vocative noun; from *serere*, "to sow") sower, planter, founder, progenitor (usually divine); originator; literally "seeder".
- **Arepo -** unknown, likely a proper name, either invented or perhaps of Egyptian origin, e.g., a coded form of the name Harpocrates or Hor-Hap (Serapis). This is a hidden word of resonance in Freemasonry like *ohm* is chanted when meditating. Now spell it backwards and we get OPERA a place of sound. This is not a word; it is a resonance but negative. Going to the opera suddenly becomes a place of worship of sound.
- **Tenet -** (Verb; from *tenere*, "to hold") he/she/it holds, keeps, comprehends, possesses, masters, preserves, sustains.
- **Opera -** (nominative, ablative, or accusative noun) work, care, aid, labour, service, effort/trouble; (from *opus*): (nominative, accusative or vocative noun) works, deeds; (ablative) with effort.
- **Rotas -** (Plural of *rota*) wheels; (verb) you (singular) turn or cause to rotate.

The church then corrupted the SATOR or ROTAS into the form of a cross, let me explain.

Here is a Greek cross (anagram) that reads *Pater-Noster* (Latin "Our Father") the opening line of the Lord's Prayer. In the square the words are placed, the cross and the crossover of Paternoster (11 letters) can be seen. Placed around the letters in 4 areas are A,O,O,A which represent *Alpha* (beginning) and *Omega* (ending), the omnipotent of God.

					P					
					A					
					T					
		A			E			O		
					R					
P	A	T	E	R	N	O	S	T	E	R
					O					
		O			S			A		
					T					
					E					
					R					

The mystery schools took the Lo Shu principles and replaced the numbers with letters and vice versa.

We all know of the Bible, but Hebrew scholars have a second scripture that has been debated to the same extent and I bet you have never heard of it, I give you the *Sefer Yetzirah*. It is the earliest scripture of Jewish Esotericism (ספר יצירה *Sēpher Yəṣîrâh*, or translated as the *Book of Formation*, or *Book of Creation*. This scripture of Jewish Mysticism while linguistic is in fact mathematical. It explores the patriarchies position of Abraham; it conveys in written form how the origins of the universe came into its present existence.

In the *Quran* there is a mention of a book (which is now considered lost) called the *Scroll of Abraham* and I believe it is about the above document.

In the British Museum the *Sefer Yetzirah* is called the *Hilkot Yetzirah* and has been declared to be esoteric lore as is not accessible to anyone but the pious. In true terms the 32 wonderous ways of wisdom brings up the mystery schools use of the number 32 as the god code number for it states the following:

"The thirty-two mysterious paths of wisdom Jah has engraved [all things], [who is] the Lord of hosts, the God of Israel, the living God, the Almighty God, He that is uplifted and exalted, He that Dwells forever, and whose Name is holy; having created His world by three [derivatives] of [the Hebrew root-word] **s^{e}f^{a}r** *: namely, sefer (a book), sefor (a count) and sippur (a story), along with ten calibrations of empty space, twenty-two letters [of the Hebrew alphabet], [of which] three are principal [letters] (i.e. א מ ש), seven are double-sounding [consonants] (i.e. בג"ד כפר"ת) and twelve are ordinary [letters] (i.e. ה ו ז ח ט י ל נ ס ע צ ק)".*

Speech is vibration and this leads to resonance. In *Stopper 1* I mention phonics which has direct relation to the words we choose to use, to make people angry or inversely make them love you. Numbers do the same. Acoustic resonance is a phenomenon in which acoustic systems amplifies sound waves whose frequency matches one of its own natural frequencies of vibration (its resonant *frequencies*).

The term 'acoustic resonance' is sometimes used to narrow mechanical resonance to the frequency range of human hearing, but since acoustics is defined in general terms and concerns vibrational waves in matter, acoustic resonance can occur at frequencies outside the range of human hearing.

These can be physically demonstrated with Cymatics sand boards.

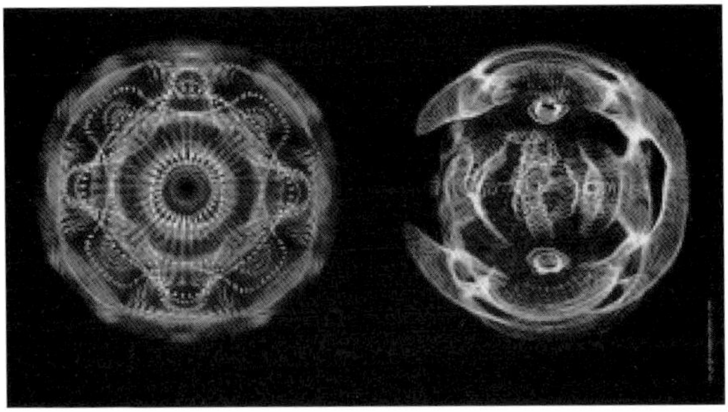

Frequency number differences can create different patterns.

The Magic squares have a direct relationship with cymatics and frequency, shaping physical forms through vibration. This is true and reversed in that numbers can create shapes which can be turned into patterns and vibration.

What is a Sigil?
It is a symbol used in magic. The term has been referred to a type of pictorial signature of a demon, angel, or other entity. It derives from the Latin *sigillum*, meaning "seal", though it may also be related to the Hebrew *segula*.

Do you remember in chapter 1 and the abracadabra pattern? What does the word mean?

A - B - R - A - C - A - D - A - B - R – A
A - B - R - A - C - A - D - A - B - R
A - B - R - A - C - A - D - A - B
A - B - R - A - C - A - D - A
A - B - R - A - C - A - D
A - B - R - A - C - A
A - B - R - A - C
A - B - R - A
A - B - R
A - B
A

A
A - B
A - B - R
A - B - R - A
A - B - R - A - C
A - B - R - A - C - A
A - B - R - A - C - A - D
A - B - R - A - C - A - D - A
A - B - R - A - C - A - D - A - B
A - B - R - A - C - A - D - A - B - R
A - B - R - A - C - A - D - A - B - R - A

It could have been derived from the equally magical word "abraxas" whose letters, in Greek numerology, add up to 365, the number of days in the year. The word is of Hebrew or Aramaic origin, being derived either from the Hebrew words *"ab"* (father), *"ben"* (son), and "*ruach ha-kodesh*" (holy spirit), or from the Aramaic *"avra kadavra"*, meaning "it will be created in my words".

It was first recorded in a Latin medical poem, *De medicina praecepta saluberrima*, by the Roman physician Quintus Serenus Sammonicus in the 2nd century AD. It is believed to have evolved into English via French and Latin from a Greek word "*abrasadabra"* (the change from *s* to *c* seems to have been through a confused transliteration of the Greek). Quintus Serenus Sammonicus said that to get well a sick person should wear an amulet (inverted pyramid) around the neck, a piece of parchment inscribed with a triangular formula derived from the word, which acts like a funnel to drive the sickness out of the body.

The mystery schools use the 'inversion method' using the pyramid upward (thereby meaning *"It will be destroyed in my words"*.

Some of these words, like 'hocus-pocus' (1634), 'abraxas' (1569) and 'hey presto' (1732), have a long history and a link to supernatural beliefs. Others, like hey-presto's American form 'presto changeo' (1905) and 'shazam' (1940) are pure stage language. 'Presto' simply means 'quickly' in Italian.

"Hocus pocus, tontus talontus, vade celeriter jubeo." It is entirely possible that this was derived from the phrase spoken at Catholic Mass: *"hoc est enim corpus meum"*, or "for this is my body".

Even the Nazi Vril Society used this expression, and can you see abracadabra in the SS Bolts? They inverted them with colour using the fashion house Hugo Boss to manufacture SS attire.

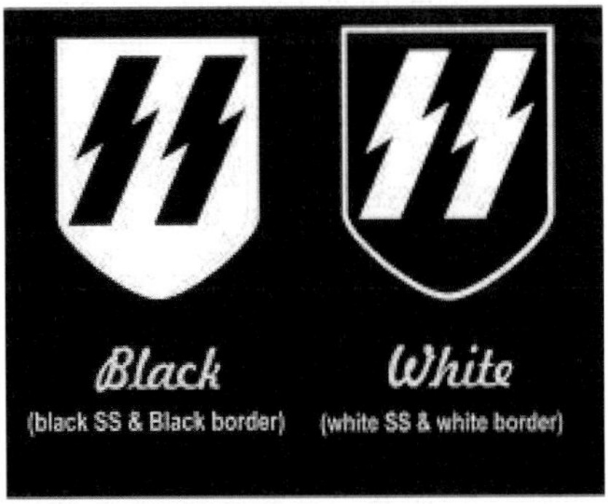

Harry Potter fans will know the Killing Curse *Avada Kedavra*, in which J K Rowling seems to have combined the Aramaic source of *"abracadabra"* with the Latin *"cadaver"*, a dead body created by the Killing Curse. *"Kadavra"* in Turkish means "cadaver" or "corpse".

1021 and 1041 Hertz.

Cymatics have a significant role in seeing how shapes can be formed from vibrational frequencies. The ancient *Grimoires* or European magic books contain chants to summon demons. These spell books concentrated on pronunciation. The science of cymatics is the study of sound vibration with matter. This can be demonstrated with sand plates and vibration which result in different geometric patterns with different frequencies. Each frequency and number create different patterns and sigils.

72 demons and Seal of Solomon

At this point you may be confused on what the Biblical Astrology, Quran and Cymatics have to do with demons? Firstly, here is a list of the 72 demons and the opposing angels from religious texts.

Angel (per Reuchlin)	Biblical verse (per Rudd)	Demon ruled (per Rudd)	Angel (per Reuchlin)	Biblical verse (per Rudd)	Demon ruled (Per Rudd)
1. Vehuiah	Psalms 3:3	Bael	37. Aniel	Psalms 80:3	Phenex
2. Ielial	Psalms 22:19	Agares	38. Haamiah	Psalms 91:9	Halphas
3. Sitael	Psalms 91:2	Vassago	39. Rehael	Psalms 30:10	Malphas
4. Elemiah	Psalms 6:4	Gamigin	40. Ieiazel	Psalms 88:14	Raum
5. Mahasiah	Psalms 34:4	Marbas	41. Hahahel	Psalms 120:2	Focalor
6. Iehahel	Psalms 9:11	Valefar	42. Michael	Psalms 121:7	Vepar

7. Achaiah	Psalms 103:8	Aamon	43. Veualiah	Psalms 88:13	Sabnock
8. Cahethel	Psalms 95:6	Barbatos	44. Ielahiah	Psalms 119:108	Shax
9. Haziel	Psalms 25:6	Paimon	45. Sealiah	Psalms 94:18	Vine
10. Aladiah	Psalms 33:22	Buer	46. Ariel	Psalms 145:9	Bifrons
11. Laviah	Psalms 18:46	Gusion	47. Asaliah	Psalms 92:5	Vual
12. Hahaiah	Psalms 110:1	Sitri	48. Mihael	Psalms 98:2	Haagenti
13. Iezalel	Psalms 98:4	Beleth	49. Vehuel	Psalms 145:3	Crocell
14. Mebahel	Psalms 9:9	Leraje	50. Daniel	Psalms 145:8	Furcas
15. Hariel	Psalms 94:22	Eligor	51. Hahasiah	Psalms 104:31	Balam
16. Hakamiah	Psalms 88:1	Zepar	52. Imamiah	Psalms 7:17	Allocer
17. Loviah	Psalms 8:9	Botis	53. Nanael	Psalms 119:75	Caim
18. Caliel	Psalms 35:24	Bathin	54. Nithael	Psalms 103:19	Murmur
19. Levuiah	Psalms 40:1	Saleos	55. Mebahaiah	Psalms 102:12	Orobas
20. Pahaliah	Psalms 120:1–2	Purson	56. Poiel	Psalms 145:14	Gremory
21. Nelchael	Psalms 31:14	Morax	57. Nemamiah	Psalms 115:11	Ose
22. Ieiaiel	Psalms 121:5	Ipos	58. Ieialel	Psalms 6:3	Auns
23. Melahel	Psalms 121:8	Aim	59. Harahel	Psalms 113:3	Orias
24. Haiviah	Psalms 33:18	Naberus	60. Mizrael	Psalms 145:17	Vapula
25. Nithhaiah	Psalms 9:1	Glasya-Labolas	61. Vmabel	Psalms 113:2	Zagan
26. Haaiah	Psalms 119:145	Bune	62. Iahhael	Psalms 119:159	Valac
27. Ierathel	Psalms 140:1	Ronove	63. Anavel	Psalms 100:2	Andras

28. Saeehiah	Psalms 71:12	Berith	64. Mehiel	Psalms 33:18	Flauros
29. Reiaiel	Psalms 54:4	Astaroth	65. Damabiah	Psalms 90:13	Andrealphus
30. Omael	Psalms 71:5	Forneus	66. Mavakel	Psalms 38:21	Cimeries
31. Lecabel	Psalms 71:16	Foras	67. Eiael	Psalms 37:4	Amduscias
32. Vasariah	Psalms 33:4	Asmodeus	68. Habuiah	Psalms 106:1	Belial
33. Iehuiah	Psalms 94:11	Gaap	69. Roehel	Psalms 16:5	Decarabia
34. Lehahiah	Psalms 131:3	Furfur	70. Iabamiah	Genesis 1:1	Seere
35. Chavakiah	Psalms 116:1	Marchosias	71. Haiaiel	Psalms 109:30	Dantalion
36. Manadel	Psalms 26:8	Stolas	72. Mumiah	Psalms 116:7	Andromalius

What do they all have in common? They each have a sigil. Do you recognise this one?

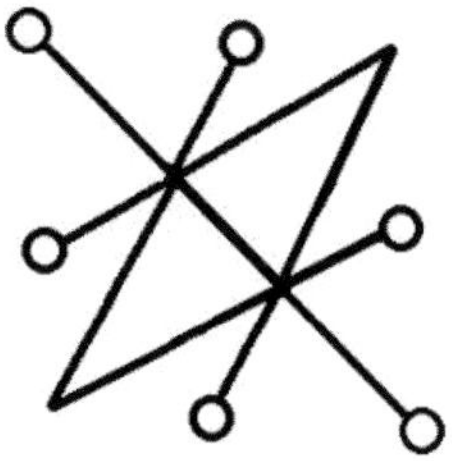

This is the symbol of AZAZEL / SAMYAZA also known as ASTAROTH who is 29 on the demon list. If you take 29 and divide by the 3 square group, you get 9.666.

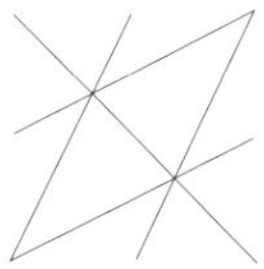

Considering the sigil or Lo Shou of Saturn, we realise that the sigil is like the one used by Aleister Crowley famous for Thelema and controversial occultists. The square has 6x6 boxes, each linear line adds up to 15, 1+5 = 6 and the 666 becomes apparent.

The magic square subject is enormous; it has computations relating to numeric calculations, resonance in frequency and sigils used by the secret societies for demonology belief. We leave this chapter with the fact that *abracadabra* is no longer a childish word.

Chapter 10 - Computers, Parachutes and Nuclear War

Computers do stuff, hard stuff, difficult stuff, and they do it fast at speeds which are called 'processing'. Some are more powerful than others and have more 'processing power'. Computers operate using '*bits*' and there are 8 bits to a *byte*. Bit is short for *Binary Digit*, the smallest unit of data in a computer. Byte is a sequence of bits, and a sequence of bytes make computer code.

Computers CPU (*Central Processing Unit*) clock speed represents how many cycles per second it can execute. Clock speed is also referred to as *clock rate*, PC frequency and CPU frequency. This is measured in gigahertz, which refers to billions of pulses per second and is abbreviated as GHz.

A PC's clock speed is an indicator of its performance and how rapidly a CPU can process data (move individual bits).

These in clock speed terms are Binary of 0s and 1s in orders of sequence to calculate a result. Binary is the primary language of computers; it is called a base-2 numeral system invented by Gottfried Leibniz. It represents electricity sequence being *allowed* to flow or *not allowed* to flow. What do I mean by sequence? If you had two light switches for the same light, one upstairs and one downstairs to light your hallway there are 4 separate ways we could flip those switches:

- Both up (off)
- First up, second down
- First down, second up
- Both down

Binary code takes each of those combinations and assigns a number to it, like this:

- Both up = 0
- First up, second down = 1
- First down, second up = 2
- Both down = 3

To write any singular letter on a computer takes 8 'lightbulbs', so a word with 5 letters would take 40 lightbulbs! How many lightbulbs do you think it took to write this page? Well, there are 364 keyboard strikes so 2912 lightbulbs in fact.

Computing speeds have got faster and faster to the point now that the desktop or laptop operate so fast, they are doing so at the speed of sound.

Things have now changed with *Quantum Computing* which uses *qubit*. $\frac{|0\rangle+|1\rangle}{\sqrt{2}}$
These quantum bits operate at the speed of light.

We are living in a universe that is operating off a binary code system of 1 and 0 or on and offs. We live in a dual reality of positive and negative energy known as chemistry, we live either masculine or feminine, an energy known as biology and we even live spiritually with a psychological energy often referred to in religion as good or evil. If we take the word good and remove an *O* we are left with GOD, add the letter D to evil, we have DEVIL. This conscious knowing of language separates humanity from the animal kingdom simply because of the size of our brain but more importantly the neuron count. The brain is a central processing unit or CPU, the more neurons the more calculations that can be made. This mental plane difference is comprehension, now look at the numbers of the animal kingdom and the approximation of neuron count:

Animal	**Approximate Neuron Cell Count**
Jellyfish	5,000
Leach	10,000
Fruit fly	250,000
Bumblebee	960,000
Mouse	71,000,000
Cat	760,000,000
Dog	2,253,000,000
Brown Bear	9,586,000,000
Chimpanzee	28,000,000,000
Human	100,000,000,000

The ability of consciousness is recognised by the mystery schools but mostly written about within the books of the Illuminati. They believe in 5 planes of existence which work as follows:

1. The Great Plain of Chemistry (atoms to cells)
2. The Great Plane of Biology (cells to animals)
3. The Great Plain of Psychology (animals to humans)
4. The Great Plane of Metaphysics (humans to angelic)
5. The Next Plane of Existence (the unknown)

My research has identified that it is the understanding of neurons which lays beneath the teachings of these mystery schools, for neurons differ from other cells in the body because neurons have *dendrites* and *axons*. The CPUs within our body have dendrites sending electrical signals and axons receiving them. The average human brain weighs 3 lbs and has a psyche, thought and an imagination element. Is it estimated that at any one time we can make 1,000,000 GB quantum computer calculations in any one instant.

Let us talk binary

01001000 01100101 01101100 01101100 01101111 says *'Hello'*. If you want to, you can write your name using the binary chart I have included in the back of this book. You now have a set of codes to use with anyone you wish in secrecy. This will be revisited in the chapter on cyphers.

Mars

Numbers are presented each day, but most people never see the occult. Hiding numbers is easy for those aware, which is why it is kept secret in the mystery schools. Binary sonogram patterns are used extensively only none-math people never see it. When I saw the chute descending from the Mars mission on the news on 18th February 2021, I did some research as I saw the binary but not the words.

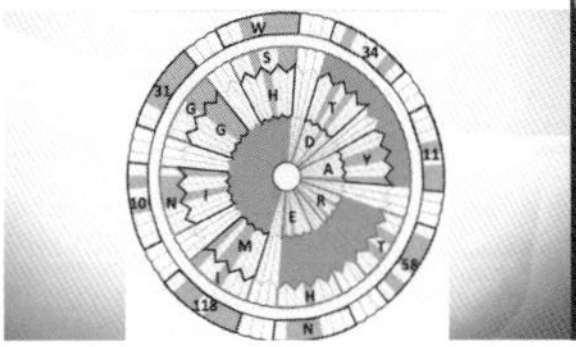

Humans spent billions on this chunk of alloy sent to space to find out if dust mites exist on the surface of Mars. Look at the pattern ring by ring and put those numbers into groups of 10, a 10-bit pattern. Each position from the right is therefore an exponent of 2, with 2 to the 0 is 1. This one is 2, then 4, 8 and so on; add up the numbers and that is what the column is. The inner 3 rings, the numbers map to letters and you get the chant of 3 "*Dare Mighty Things*". The hidden binary message was from a speech in 1899 by United States President Theodore Roosevelt.

The pattern on the outer edge represents the GPS coordinates: 34.2013° N, 118.1714° W which on Google GPS positioning turns out to be *the Jet Propulsion Laboratory* in Pasadena California.

These NASA boffins spent thousands of dollars on hidden codes that I worked out in 3 hours. I have sent them a penned letter stating it may be a waste of money as I do believe the dust mites on Mars cannot read, I never gotten a written response.

Enigma machine

Alan Turing, the father of the computer and artificial intelligence was another genius and understood programs, programming and did so using only slide rules. In 2001 the U.K. Bank of England released the new £50 note.

Look closely at the numbers and take his full name – 'Alan Matheson Turing OBE' first. The binary on the note is – 1010111111110010110011000 and it really means 2306. Alan Turing's birth date was 23rd June 1912 or 23rd of the 6th, they had fun with this one.

Alan's comment about computers and state on the note is *"This is only a foretaste of what is to come and only the shadow of what is going to be"*. He foresaw the potential of what computers could be capable of doing.

DNA and Demons

In *Stopper 1* conclusion, I wrote about the word Demon root word origin Greek DAIMON meaning 'wise'. In fact, the true origin when researched deeper is genius, in 1867 the Physicist James Clerk Maxwell wrote a hypothesis of a (idea) thought experiment describing the violation of the second law of thermodynamics. The basis of this idea was to create a physical mechanical engine that creates more output, then input more energy for less fuel for example.

This law of thermodynamics proved that more energy always goes into a thing to create energy out of a thing. My investigations into this work found he used the expression DEMON as a supernatural being working in the background rather than eminent malevolent one. The basis of his argument I understood immediately as I have installed many ground source heat pumps, these are simply boxes that separate hot air from cold air using a pump for either cooling or heating a building. While these pumps are efficient at heating environments, the electricity input for the pump can become the coefficient of more energy being put in. For pumps to work they need electricity, electricity creates heat and heat gets wasted. Too big a pump and it becomes inefficient monetarily. Replacing the pump with Maxwell's idea of a Daemon envisages the Demon opening a trap door letting cold molecules one-way and hot molecules the other.

Tesla's CEO Elon Musk on 24th October 2014 warned about Artificial Intelligence (AI) which he called *"our biggest existential threat"*, he believes there should be some regulatory oversight at both national and international levels. This isn't the first time he mentioned it, in August he tweeted that *"We need to be super careful of AI. Potentially more dangerous than nukes"*.

Speaking at the MIT AeroAstro Centennial Symposium he added that with AI, *"we are summoning the demon"* and continued *"In all those stories where there's the guy with the pentagram and the holy water, it's like yeah, he's sure he can control the demon. Didn't work out"*.

The speech was fascinating because he also added *"if you want a picture of A.I. gone wrong, don't imagine marching humanoid robots with glowing red eyes imagine tiny invisible synthetic bacteria made of diamond, with tiny on-board computers, hiding inside your bloodstream and everyone else's. And then, simultaneously, they release 1 µg of botulinum toxin, and everyone just falls over dead"*. This friendly little discussion about the future of humanity scared me. Inventors invent, humans discovered fire only when things burned down and then we create the fire extinguisher. Then the fast car is followed with us creating breaks to stop accidents. Seatbelts followed thousands of deaths as an afterthought.

Humans are linking AI with nuclear weapons and that is a lesson we cannot find a solution should the computer decide to "let's play a game". In the movie *WarGames* (1983) a computer plays tic-tac-toe or noughts and crosses (both the chant of 3). The computer was called War Operation Plan Response – WOPR (pronounced WHOPPER), the supercomputer at NORAD in the USA. It ran war simulations and learnt over time there could be no winner for humankind.

D-Wave

In mathematics, humans are in their infancy as we need calculators (glacial speed), we have had computers (speed of sound) and now we are using quantum computers which use the speed of light. Objects can be coded into information and physically created by 3 dimensional printers.

It is now upon us that computers will govern things for humanity and I myself have moved away from my smart phone and only use it on specific occasions. I have a concept that in the future if your mobile phone realises that you are about to replace/upgrade it and you suddenly need to call an ambulance it might not do it as it is not in phone's best interest. Computers while using binary are now thinking for themselves. These are called D-wave computers and the D stands for Demon.

"So if you're standing you have the opportunity to stand next to one of these machines, it is an all inspiring thing at least for me. It feels like an altar to an alien God. They really are impressive machines" Dr Geordie Rose founder and CTO at D-Wave Computers.

Bitwise Negation is the slow speed of sound, binary speed of sound and here is its mathematical pyramid.

0
0 1
0 1 1 0
0 1 1 0 1 0 0 1
0 1 1 0 1 0 0 1 1 0 0 1 0 1 1 0

Super computers however have an error number which is, 100000000000000000000000001.

Belphegor's Prime Number

This palindromic number 1,000,000,000,000,066,600,000,000,000,001 ($10^{30}+666\times10^{14}+1$), is a number which reads the same both backwards and forwards and is only divisible by itself and 1. This was discovered by Harvey Dubner; the name Belphegor refers to one of the *Seven Princes of Hell* who was charged with helping people make *ingenious inventions and discoveries*, "Belphegor's prime". The number itself contains superstitious elements that have given it its name: the number 666. Used in symbolism by the mystery schools to represent one of the creatures in the Apocalypse or, more commonly, the Devil. This number is surrounded on either side by thirteen zeroes and is 31 digits in length (13 reversed), with 13 itself long regarded superstitiously as an unlucky number in Western culture.

It contains two sets of 13 zeros with 666 in the centre.

31
1000000000000066600000000000001
13x0   13x0
666

If you remove all the zeros you are left with 16661, so appears the year 1666.

Binary has been placed on the largest bank note in England and it will be overtaken by crypto. The D-wave will sit behind it which uses Belphegor's prime, I think you can see the correlation issue.

Chapter 11 - Music, Pyramids and Tesla's 3, 6 and 9

Sir Arthur C Clarke was an English science fiction-writer, futurist, inventor, undersea explorer, and television series host. He died on the 19th March 2008; his most famous screenplay was *2001: A Space Odyssey*. I have read many of his works and there is one statement he made that has resonated with me:

"Any sufficiently advanced technology is indistinguishable from magic"

This statement explains that if you did not know any advanced technology you would regard it as something spiritual, otherworldly, or magic. Taking an innovative technology that nobody else understands and when used in a physical manifestation you can utilise it for either good or evil.

Sound

Numbers are in music; that we can agree, look at words commonly used in discussion that are related to music:

- To *conduct* yourself better; preaching to the *choir*; blowing your own *trumpet*; *sound* like a broken record; *music* to my ears; change your *tune*; it strikes a *chord*; *singing* from the same song sheet; stop *harp*-ing on about it; all that *jazz*; *playing* second fiddle; *beat* the band; *beat* a path; all the *bells* and *whistles*; march to the *beat* of your own *drum*; fit as a *fiddle*; the same old *song* and dance; it's not my *forte*; *play* it by ear; it *rings* a bell; it ain't over till the fat lady *sings*; it strikes the right *note*; *sounds* about right. Even the human body is made of *organs*, it is music.

The 3 tritones of enrichment are the *octave 2:1, the fifth 3:2* and the *fourth 4:3*. If you do not know what this means, let me explain. Try singing aloud on your own without either your pet or partner present. If you are not alone, then I take no responsibility for the ensuing crises, evacuation, or call to the relevant authorities to have you committed.

The first 2 notes of each song are my way of getting you tuned in, so first.

- Octave 2:1 is the first 2 notes of *"Somewhere over the rainbow"* off you go sing it, oh what fun we are having.
- Fifth 3:2 can best be remembered with the first 2 notes of *"Twinkle Twinkle Little Star"*, come on...I had to sing goodness knows how many songs to figure this out ,so join in please...has the cat left yet?
- Fourth 4:3 (not augmented fourth as in the *stopper 1*) is best remembered with the song *"Here Comes the Bride"*, once you start you cannot stop, I know...and your partner now has serious concerns about your mental health. These 3 double notes are said to have health benefits, but what lays within the numbers.

At this point I will also reveal to you that when humans sing in a choir their heartbeats synchronise in rhythm, they become a collective.

The string lengths when played have ratios 2 to 1, 3 to 2 and 4 to 3, respectively. Harmonious proof that there is an order in sound but this natural order in numbers exists everywhere. These ratios are in everything and easily researched.

- Water holds 2:1 or 2 atoms to 1 in water H_2O
- Mercury holds a 3:2 ratio as it rotates 3 times, it orbits the Sun 2 times
- Televisions and cinema use to run a 4:3 ratio as it was to capture the ideal speed for the human eye

What is frequency

When you strike a tuning folk, the note will resonate to the frequency to which it has been tuned. If you take 2 of them and strike 1 but hold them close, the 1st resonates the 2nd and both ring at the same frequency. 1st will active the 2nd reaction. This is where the term "*he is on my frequency*" stems from.

Studies have shown that cancer cells are vulnerable to the frequencies of 100,000Hz and 300,000Hz. Frequencies and sound make a physical presence in the universe. Chladni plates and sand when vibrated at certain frequencies show the *'Cymatics'* or what sound looks like as mentioned in the magic squares chapter. Sound physically manifests form.

Healing frequencies

The *Tablet of Shamash* is a stele which was recovered from the ancient Babylon city of Sippar in southern Iraq in 1881, it is a stone object which accurately depicts frequencies in detail. The ancients knew far more about frequency then we are told. Here are some of these frequencies:

40Hz

Flickering lights and sounds at 40 hertz have been used in Alzheimer's therapy studies to stimulate an increased neural response and fight symptoms of dementia. It has been linked to gamma brain waves and the stimulation of memory.

174Hz

A Solfeggio frequency, it is a series of tones used in sacred music and used in alternative medicine which has different positive effects on human health. It is associated with the reduction of both pain and stress.

285Hz

This is also one of the Solfeggio frequencies and is considered instrumental in the healing of cuts, burns, and other physical wounds. These sound frequencies are supposed to activate the body into cellular regeneration, encouraging it to heal itself in the event of an injury.

396Hz

This sound frequency is associated with the removal of fear and other negative feelings. As one of the Solfeggio frequencies, the tone aids in the removal of guilt making it an effective addition to spiritual music. The frequency balances the root chakra while simultaneously transforming negative emotions such as grief into positive, joyful ones.

417Hz
Instead of a focus on physical ailments, healing sound therapy involving 417Hz (another of the Solfeggio frequencies) helps remove energy surrounding a past trauma or negative energies in the ambient environment. This therapy is designed to dissolve emotional blockages and activate the chakras according to these practitioners.

432Hz
This therapy is aimed at the heart chakra and listening to this frequency is supposed to lead to greater levels of mental and emotional clarity. The tuning at this frequency is considered optimal for singers and is associated with a higher level of spiritual development. Often referred to as the Miracle Tone of Nature.

440Hz
Music that is tuned to the 440Hz v 432Hz is considered mind music that aids in the listener's cognitive development. These sound frequencies are considered to activate the third eye chakra. I visited this in *stopper 1*.

528Hz
Also known as the love frequency, it is one of the most well-known and popular of the Solfeggio frequencies. This musical tone is also known as the miracle note and has been used by native populations as a sound associated with blessings since before written history. It is also said to repair DNA.

639Hz
It is said to balance the heart chakra. This sound frequency is associated with therapy intended to produce positive feelings and greater connection to harmonious interpersonal relationships. As therapy, exposure is said to encourages clearer communication practices and situational awareness. Often called the Relationship Frequency.

852Hz
This is a tone that is associated with redirecting the mind away from overthinking, intrusive thoughts, and negative thought patterns. These patterns of thought play significant role in depression and anxiety. Exposure to this sound frequency is thought to help alleviate the role of negative thoughts or psychological ailments.

963Hz
These frequencies are associated with the activation of the Pineal Gland and higher spiritual development. This frequency is known as both the Frequency of Gods and the Pure Miracle Tone. It is associated with the activation of the crown chakra and a connection to the source of all humanity.

DNA is composed of a liquid crystalline substance that acts as a form of antenna, receiver, and transmitter of holographic information impacted by subtle waves of energy. It would seem these frequencies redirect or correct these crystalline substances. For health benefits, people use devices such as *rife* machines which set the frequencies so the body can experience them. Royal Raymond Rife was born on 16th May 1888 and was an American inventor of both the electron microscope (magnification) and frequencies with multiple versions of his *rife* machines. Rife thought bacteria were behind many diseases like cancer. His research through his microscope showed that you can spot them, and his rife frequency generator was thought to be able to destroy them using electrical pulses. The American Medical Association condemned all Rife's work regarding the machine but not the electron microscope.

Words are spoken with phonetics and phonics or sound which in effect is a frequency the same can be said with numbers. It seems that this is what secret societies have held back from humanity. Recent frequencies that have been included in modern patents are (by patent date):

- 1976-Patent No. US3951134A - Apparatus and Method for Remotely Monitoring and Altering Brain Waves (MK-Ultra) (Filed 1974).
- 1983-Patent No. US4395600A - Auditory Subliminal Message System and Method (Filed 1980).
- 1989-Patent No. US4858612A - Hearing Device. A method and apparatus for simulation of hearing by introduction of a plurality of microwaves (Filed 1983).
- 1988-Patient No. DE3628420A1 - Device for reproducing voice information in subliminal technique (Filed 1986).
- 1989-Patent No. US4877027A – Hearing System. This relates to multiple levels of hearing systems all included within one patent (Filed 1988).
- 1993-Patent No. US5211129A – Syringe-implantable Identification Transponder. Is a sufficiently miniaturised to be syringe-implantable, thus avoiding surgical procedures (Filed 1991).
- 1999-Patent No. US5935054A - Magnetic Excitation of Sensory Resonances. Influencing the nervous system (Filed 1995).
- 2003-Patent No. US6587729B2 – Apparatus for Audibly Communicating Speech Using the Radio Frequency Hearing Effect. This is for communicating speech using radio frequency hearing affect to the human mind can audibly pick up on the words (Filed in 1996).
- 2000-Patent No. US6039688A – Therapeutic Behaviour Modification Program, Compliance Monitoring and Feedback System. The system enables development of a therapeutic behaviour modification program (Filed 1997).
- 2000-Patent No. US6011991A – Communication System and Method including Brain Wave Analysis and/or Use of Brain Activity (Filed 1998).
- 2004-Patent No. DE10253433A1 – Thought transmission unit sends modulated electromagnetic wave beams to human receiver to influence thoughts and actions without electronic receiver (Filed 2002).

Light as Frequency

Dr Peter P Garyaev, Russian scientist and a wave genetics specialist was nominated for a Nobel Prize on Medicine a few weeks before his death from a brain oedema. He carried out experiments on frog eggs with lasers, no big deal until you explore what he did. He charged the lasers with energy passing the beam through salamander DNA, then passed the light through frog eggs with the wave pattern. The result was not cooked frog DNA but eggs which later hatched, not as frogs but as salamanders. The authorities or watchdogs of orthodoxy panicked (Lysenkoism) and are trying to shut down this wave genetic technology. If you can imagine gene loaded laser directed at to an organ in distress without surgery to organically create a repair, the health care industry would change overnight and for ever. Wave genetics would allow humans to live far beyond our present understanding and the frequency number is 584 Hertz.

This year (2022), a man whose work I have been closely following died. His name was Luc Montagnier, a Nobel Prize winning French virologist. In 2011 he took 2 hermetically sealed tubes containing water, the 1st holding trace amounts of DNA and the 2nd water, he then electrified both tubes at a 7 Hertz EM field. The scary result was that the tube holding no DNA rearranged itself into holding DNA with a standard PCR test...not an exaggerated one. The implications are far reaching, if DNA can reproduce in electric signals, it means DNA can teleport through space and time. Science is calling it heretical, similarly like the church once called protestants. His subsequent death has stalled all research.

This then begs the question, if no life exists in water and at 7 Hertz it can move, be copied, or ghost itself in presence, then DNA can work on a quantum field effect. Darwinian evolution would shatter under such a confirmation and life could be created spontaneously from another dimension. "Beam me up Scotty" is possible it would seem.

Try googling the following patent number US6506148B2... off you go. This is a patent filed in 2001 and published in 2003 to an invention that relates to the stimulation of the human nervous system. Using an electromagnetic field at 10Hz alternating voltage of 400v it was painted under number 5169380. This patent is about the manipulation of the nervous system by pulsing images displayed on a computer monitor or TV set. The image pulsing may have embedded material or overlaid by modulating video streams either as an RF signal or video signal therefore manipulating the mind of the individual watching. Turn the dammed thing off or better still throw it away like I did.

Associate Professor, Director of Music Technology at Skidmore College, Anthony Holland, told an audience at a Ted talk on the 22nd Dec 2013 that he has a dream. That dream was to see a future where children no longer suffer from cancer, the effects of toxic cancer drugs or radiation treatment. He and his team believe they have found the answer. Have you ever heard of people shattering glass with the sound of their voice?

These researchers wondered if they could induce the same effect in a living micro-organism or detrimental cell. From ovarian, pancreatic, and leukaemia cancer cells Holland hopes that one day the treatment will override the toxic conventional treatments currently using the frequency 111Hz using the 11th harmonic.

In 2020, David Mittelstein a Biomedical Engineer at the California Institute of Technology, in Pasadena using low-intensity ultrasound, target cancer cells based on their unique physical and structural properties. Any spill over of the energy should cause little harm to healthy tissue. The treatment sends out pulses of sound waves (energy) that have a frequency above 20,000 hertz (cycles per second), which is too high for our ears to hear (that is also what makes it "ultra" sound). Medical imaging relies on noticeably short pulses of this low-intensity ultrasound. 444Hz is the frequency at the time of authoring this book, that they were working on. It has yet to be tested on humans.

We can see that frequency and vibration clearly exists in the universe and humankind may be using them in secrecy for detrimental reasons. Let us investigate antiquity and discuss their understanding of these mathematical patterns.

Pyramids

The Great Pyramid of Giza has frequency in patterns, look at its general size and shape:

- The half perimeter (two sides) is the biggest horizontal visible side
- Its hight is its biggest vertical visible dimension
- Internally there is a structural harmony
- The full height divided by 2 gives the chevroned summit in the upper chamber
- The full height divided by 3 gives you the height of the upper chamber ceiling
- The full height divided by 4 gives you the chevroned summit in the middle chamber
- The full height divided by 5 gives you the floor in the subterranean lowest chamber
- There is no 6th
- And by 7 the floor in the middle chamber
- Its 4 faces divided by its base and achieves the number Pi, π or 3.14

I believe the pyramids were not just a tomb, so let me explain.

The location of the pyramids in the Giza Plateau needs to be numerically pointed out now as:

29.97.92 N the speed of light travels at 299,792,45 8 m/s
31.13.13 E

They are at latitude 30 or 1/3rd of the way between the Equator and the North Pole. I should point out that there is a small discrepancy of the latitude 30 today due to the constellation years being travelled by the Earth, but not so when the pyramids were built.

At 481 feet high, its base covers 13 acres and weighs 6 million tons. It is closely aligned to the four geographic cardinal directions (not magnetic) North, East, West, and South with a deviation of around 3 minutes and 38 seconds of Arc, which even in today's building terms is minuscule.

The pyramids are made of circa 2-2.5 million blocks; if we worked 365 days a year on a 12-hour sunlight day, for 20 years none stop we would have to lay a stone every 2.3 minutes. As a builder I am now worried about my tea breaks!

The base and height reveal numbers that are in another chapter of this book, at a scale of 1:43,200 planet Earth fits inside perfectly. If you multiply the base perimeter of the pyramid 3024x43200 you get the equatorial circumference of Earth. Then if you take the angles.

- Half of that is 45° or 4+5 = 9
- Half again is 22.5 or 2+2+5 = 9
- Half of that is 11.25 or 1+1+2+5 = 9
- Half again 5.625 or 5+6+2+5 = 9
- The angle of the sides is 51° or 5+1 = 6
- There are 3 sides to each triangle
- It has 4 sides and 3x4 = 12 and 1+2 = 3. Each side has 2 triangles within them so that makes a total of 8 sides, then if you include the base you end up with 9

- Take the inner angles of a triangle a+b+c = 180° and 1+8+0 = 9
- Add up all base numbers 1+2+3+4+5+6+7+8+9 = 36 or 3,6 and 3+6 = 9, they are all Tesla numbers.

Tesla

Nikola Tesla was a Serbian-American, an inventor, electrical and mechanical engineer (1856–1943) and like me had an interest in the occult but his forte was frequency. His inventions included an induction motor that ran on alternating current (AC) and he also touched on ideas that were developed by others, like x-rays, neon lights, and remote controls.

"The day science begins to study non-physical phenomena, it will make more progress in one decade then in all the previous centuries of its existence" - Nikola Tesla.

Tesla suffered from OCD (chant of 3). Obsessive-compulsive-disorder adds up to 27 letters in total 2+7 = 9 and in Gematria it is:
15+2+19+5+19+19+9+22+5+3+15+13+16+21+12+19+9+22+5+4+9+19+15+18+4+5+18 = 342 which is 3+4+2 = 9.

What is all this 3, 6 and 9

Another Tesla quote: *"If you want to find the secrets of the universe, think in terms of energy, frequency and vibration"*. He was also alleged to have said *"If you only knew the magnificence of the 3, 6, and 9, then you would have a key to the universe"*.

He was so obsessed with these 3 numbers that he engaged in some compulsive behaviours around them. Walking around the block 3 times before entering the building, choosing only hotel room with the number that was divisible by 3, washing his dishes with 18 napkins (18 is 1+8 = 9, it is also divisible by 3, 6 and 9) and he ate with 6 napkins.

If you watch YouTube videos at an increased clock setting, then you are in a world of 3, 6 and 9. Here are the speeds and the numbers they calculate too:

2160p	add the numbers	2+1+6=9
1440p		1+4+4=9
1080p		1+0+8=9
720p		7+2=9
360p		3+6=9
240p		2+4=6

Write out 1-9 horizontally with vertical formation:

1	2	3	456789
9	8	7	654321
1	2	3	456789
9	8	7	654321
1	2	3	456789
9	8	7	654321
1	2	3	456789
9	8	7	654321
1	2	3	456789
41	42	43	Look at this last column in vertical terms, can you see the patterns?

Fibonacci, Telsa and 3,6,9

The code is simple in nature and the calculation works like this:

- 1+1 = 2 Take the last number before the = which is 1, add to the number that follows the = in this case 2
- 1+2 = 3 Then 2+3 = 5
- 3+5 = 8 As above, same method and you get 5+8 = 13
- Keep adding the two numbers on either side of the = sign, which runs to infinity; 0,1,1,2,3,5,8,13,21,34,55,89 and know you know the first few numbers of the Fibonacci sequence. As a 12th century mathematician his arbitrary sequence runs everywhere. Pinecones, flowers, shells, rams' horns, weather storms, spiral galaxies, it is in everything and hidden within it is another important number in mathematics called the golden ratio which is denoted by the Greek letter phi.

This is known to the mystery schools but let us now look at other patterns.

There are boasts that Tesla may have caused the 3 earthquakes in 1899 in Cape Yakataga and Yakutat Bay (Alaska) from his laboratory in Colorado Springs. Here are some numerical values behind the dates the 3 earthquakes that took place:

September 3rd, 1899	or	Sept = 9, 3rd = 3, 1899 is 1+8+9+9=9 (9,3,9)
September 6th, 1899	or	Sept = 9, 6th = 6, 1899 is 1+8+9+9=9 (9,6,9)
September 9th, 1899	or	Sept = 9, 9th = 9, 1899 is 1+8+9+9=9 (9,9,9)

We have looked at shapes, but let me refresh your mind:

- A circle has 360° or 3+6+0 = 9
- Half a circle has 180° or 1+8+0 = 9
- Quarter of a circle has 90° or 9+0 = 9
- 360° circle divided by 2 = 180 or 1+8 = 9
- Half of 180 = 90 or 9+0 = 9
- Half that = 45 or 4+5 = 9
- Half that = 22.5 or 2+2+5 = 9
- Half that = 11.25 or 1+1+2+5 = 9
- Half that = 5.625 or 5+6+2+5 = 18 and 1+8 = 9

- Half that = 2.8125 or 2+8+1+2+5 = 18 and 1+8 = 9
- Half that = 1.40625 or 1+4+0+6+2+5 = 18 and 1+8 = 9
- And on and on...

3

The pattern within 3:

3x1 = 3		3
3x2 = 6		6
3x3 = 9		9
3x4 = 12	1+2 = 3	3
3x5 = 15	1+5 = 6	6
3x6 = 18	1+8 = 9	9
3x7 = 21	2+1 = 3	3
3x8 = 24	2+4 = 6	6
3x9 = 27	2+7 = 9	9

3 hides the 3, 6 and 9 code in the final vertical column

6

The pattern within 6:

6x1 = 6		6
6x2 = 12	1+2 = 3	3
6x3 = 18	1+8 = 9	9
6x4 = 24	2+4 = 6	6
6x5 = 30	3+0 = 3	3
6x6 = 36	3+6 = 9	9
6x7 = 42	4+2 = 6	6
6x8 = 48	4+8 = 12	1+2 = 3
6x9 = 54	5+4 = 9	9

6 also hides the 3, 6 and 9 code in the final vertical column

And finally, 9

9x1 = 9		9
9x2 = 18	1+8 = 9	9
9x3 = 27	2+7 = 9	9
9x4 = 36	3+6 = 9	9
9x5 = 45	4+5 = 9	9
9x6 = 54	5+4 = 9	9
9x7 = 63	6+3 = 9	9
9x8 = 72	7+2 = 9	9
9x9 = 81	8+1 = 9	9

In science as K is constant, so is number 9 for Tesla

In 1900 the US-Patent 787,412 titled *"Art of Transmitting Electrical Energy Through the Natural Medium"* was filed by Tesla and issued in 1905. The devise he was patenting was later to become known as the *Electromagnetic Pyramid*.

9 is hidden within the pyramids but here are a few others:

- Pyramid base circumference is 43200 4+3+2+0+0 = 9
- The base divides by 600 6+0+0 = 6
- Latitude of the Great Pyramid is .2 off 30 3+0 = 3
- One zodiac age is 2160 years 2+1+6+0 = 9
- One Great Year is 25920 long years 2+5+9+2+0 = 19 1+8 = 9

Numbers not in the pyramid:

- 1+1 = 2
- 2+2 = 4
- 4+4 = 8
- 8+8 = 16 and 1+6 = 7
- 16+16 = 32 and 3+2 = 5
- 32+32 = 64 and 6+4 = 10 and 1+0 = 1
- Did you notice that within this pattern there is no 3, 6 or 9, just 1,2,4,5,7, and 8

When studying Tesla's work, I concentrated on 3, 6 and 9 but upon reflection when I looked at the remaining numbers, I found something extraordinary. By using the other numbers from 1 to 10 leaving out 3,6,9 a pattern emerged. Start with 1 then double to 2, double to 4, you then see repetitions appear of 1,2,4,8,7,5 which repeats on the right-hand side.

1=1
2=2
4=4
8=8
16=7
32=5
64=1
128=2
256=4
512=8
1024=7
2048=5
4096=1
8192=2
16384=4
32768=8
65536=7
131072=5

This shows that Tesla's throwaway numbers indeed have a code. Do you see it?
1+2+4+8+7+5 = 27 and 2+7 = 9

Let us go back to frequencies briefly at this point.

40Hz divided by 12 (months in a year) = 3.333333333
396Hz divided by 12 = 33
432Hz divided by 12 = 36

440Hz (the magical music number) divided by 12 = 36.66666666
432Hz (which is known as Verdi 'A') divided by 12 = 36 or 3 and 6. If you delve into this number you will notice that 432 squared calculates to the speed of light to 1% accuracy
639Hz Tesla's numbers
963Hz also Tesla's numbers

Here is another occult pattern which hides in 12 (months) or 1+2 = 3:

1x12 = 12	1+2 = 3		3
2x12 = 24	2+4 = 6		6
3x12 = 36	3+6 = 9		9
4x12 = 48	4+8 = 12	1+2 = 3	3
5x12 = 60	6+0 = 6		6
6x12 = 72	7+2 = 9		9
7x12 = 84	8+4 = 12	1+2 = 3	3
8x12 = 96	9+6 = 15	1+5 = 6	6
9x12 = 108	1+8 = 9		9

3+3 = 6 and 6+3 = 9 and 3+6+9 = 18 and 1+8 = 9 they truly are a remarkable 3 numbers.

Theoretically my 1st book should never have sold but I placed what I see as Tesla energy into my work. I sold the book on (as some have said 'awful') Amazon for 19.99, I discounted the book directly by 3.33 giving you 16.66. These are Tesla's numbers, given the book reached 54 countries there seems to be something energetic within these numbers.

Marconi patents were overturned in favour of Tesla in 1943 but as he was dead the government took possession. Interestingly the US Supreme Court decision number was (you guessed it) 369. Another coincidence?

As we end this chapter, be aware that frequency knowledge is being suppressed and this is worth debating. I believe the purpose of the Great Pyramid of Giza was for spiritual initiation and healing, but it could be a balancing instrument for the Earth's energy grid.

Tesla believed the Earth was a giant free energy generator. Rife believed frequency could extend life. Both had their life's work destroyed by governments and they died in poverty. In this chapter I have collated the important numbers for others to interpret it as they see fit. The one thing I am sure is that the pyramids are not just a burial chamber, they are communicating numbers of time, distance, and frequency to humanity. I leave you with this thought; could a choir whose hearts are synchronised, *sing* people into health? and are resonance chambers the next leap to longevity, leaving the pharmaceutical industry bankrupt? Why do you suppose that when you hear a tune in the morning, you tend to sing it for the rest of the day? Could it be that it has you tuned in, *'singing yourself into health'?*
The frequencies for attacking cancer cells seems to be 111 and 444 Hz. 1+1+1 = 3 and 4+4+4 = 12 and 1+2 = 3.

Chapter – 12 - Body Parts, Distance and the Soul

There are many words that describe parts of the human body which laypeople do not know; they are used by the medical profession:

- The space between your eyebrows it is called the *glabella*
- The rumbling of the stomach is called a *wamble*
- The cry of a new-born baby is called the *vagitus*
- Your little finger or toe is called *minimus*
- The space between your nostrils is called *columella nasi*
- The utterly sick feeling you get after eating or drinking far too much is called *crapulence*
- *Foramen magnum* Latin root word 'really big hole' or the cavity of the skull
- *Hippocampus* the part of the brain that is involved with memory and learning, comes from the Greek meaning 'seahorse'
- *Lumbricals* the inner sinews connected to the fingers in the hand and is from the Latin 'earth worms'
- The upper part and the largest region of the hip bone is also known as the *ilium*, it is also known as the Girdle of the Sky (same as the zodiac)
- Solar plexus or sunlight in your stomach

A human baby is born with 300 bones and with the growth of the sternum, the joining of skull plates, and cartilage fusing into bone in the axial skeleton into adulthood results in 206.

As well as Latin origin others are from the Greco / Roman writings and leave the Phoenician and Arcadian history a little vaguer. One thing regarding measurement which will become obvious is that a human's measurement is geometric, even with the planets and distance.

Man is the measure of all things

Humanity had to start measuring in order to build, sow seeds, or walk distances. Measurement while not natural in a human's mind became ever more essential in the evolvement of our species.

Ever heard the term "*rule of thumb*" (chant of 3)? Its origins are right in front of you, but the occulting of these facts has been removed as late as 100 years ago.

Compared to the ancients we consider ourselves in a higher technological state, this however in my opinion is pure ignorance. All design depends on symmetry not only from nature but universal balance. A well know architect stated, "*form follows function*"; the more clarity of form in design suits mankind's actual need and is called *ergonomics*. A 70's style chair may be attractive to the eye but sitting on it for hours on end can be torture.

Here are some measurements you may have heard of but not necessarily the description set below:

- An *inch* is the width of an adult male thumb.
- A *foot* is 12 thumbs width.
- An *English foot* is 3 palm widths (each is 4 fingers width) or 16 fingers.

- An *Egyptian cubit* is the distance between your elbow to the middle finger (7 palms or 28 fingers).

Now here is the deception, a 'foot' is NOT based on feet but on thumbs, fingers, hands, and normal cubits. The Egyptian royal cubit is the earliest attested standard measure, it is 0.52359 metres. 6 modern feet is 4 Egyptian cubits and that is 1 fathom.

Fathom originally was a measurement of the distance between your outstretched arms in a crucifix stance, the distance is measured from middle finger to middle finger which is where we get the term (*"to fathom out your own existence"*). Sailors used weighted rope thrown into water to fathom the depth, when drawing up the rope they measured 1 fathom by outstretching the rope and counting each single fathom at a time. Interestingly your fingertip-to-fingertip fathom is also your height.

Half of a fathom is 1 yard. Historically it was called a yard stick (*"a gauge of truth / measurement of truth"*). To fathom something is to understand, a yard stick is a measure of truth.

A mile

Simply put it is 1 thousand human paces but not steps. If you take a pace forward, starting off with your left heel on the ground landing on your right heel this is not the measurement, it is when your 2nd pace and left heel hits the ground again is when you count as you move forward and count every other step. 2 actual steps count as 1. A mile equates to 5280 feet in distance, so 1 thousand of this measurement 1000 x 528 = 5,280 feet, or mile as 1 thousand human paces.

Furlong

Mentioned elsewhere in geometry, consider 12,000 paces = 1 furlong or 660.001 feet or 7,920 inches, the exact diameter number of the Earth at 7,920 miles.

The pace to the holy city of Jerusalem has a direct correlation to a human foot and the size of Earth. It is a fact that humans are designed in size with direct measurements to the Earth. This is worked out as follows:

$$\frac{Pace}{Foot} = \frac{Earth}{Holy\ City} = \frac{7.920}{1{,}500} = \frac{1}{5.28}$$

The Soul has Weight

The church considers the human body as 3 distinct things: *Body, Soul* and *Spirit* and this has a collective name called *Trichotomism*. In Genesis 2:7 *"And the Lord God formed man from the dust of the ground and breathed into his nostrils the breath of life; and man became a living soul"*.

In 1907 Duncan MacDougall, a physician from Massachusetts USA produced a paper claiming he measured the weights of 6 patients at the point of death who were paralysed with various existing illnesses. Three patients were found to lose 3 quarters of a once or 21.3 grams. Hotly discussed in the medical profession, dissenters blamed this on sweat leaving the body as the lungs stopped cooling. However, the problem is that this sudden released was in seconds not minutes. Interestingly dogs were experimented on in another series of tests and upon death no weight loss was found to take place. The '21-gram experiment' as it became known was banned but the question remains.

If humanity does have souls and they are 21.3 grams, the air we breath is filled with conscious energy you need to consider this with a *Deep Breath* and 2+1+3 = is another 6.

Cells

Celestial bodies are cells, planets are cells, humans are communities, and humans are a collection of cells. Every living cell has a voltage, on the inside is negative and on the outside is positive. The cells in a human body holds circa 0.7 volts with an estimated average of 50 trillion cells in the human body. Look at *stopper 1* and read pages 78 – 79 before moving on. Consider 0.7 volts multiplied by 50 trillion cells and we get 3 trillion 500 billion volts of electricity in your body at this very moment. All this energy creates vibrations with some positive and others negative, ever heard the saying *"he gives off bad vibes"*? it means vibrations.

Every single cell of the human body is 70 millivolts of electricity (.07 volts), and the human body contains around 37.2 trillion cells. When you extrapolate that out it means you have more than 2 trillion, 604 billion volts of electricity and you are a walking mini nuclear reactor.

It is at this point I require you to ponder a thought. We are told to trust our feelings and numbers have no feelings, well they do...and do I hear you question where I am going this? Humans beings are an energy wave that happens to believe that we are only physical. One individual can affect another with negative waves or *"vibes that are bringing me down man"*.

Quantum physics does not study the physical particles, it is the study of energy waves. Is your body an illusion?

Quantum physics takes the waves and adds them together, collectively they are called a *field*. Electroencephalography machines (EEG) interpret electrical brain waves from small metal disks with wires pasted onto the scalp but not field energy because all thoughts of the human mind are inside the head, right? wrong. I introduce to you the Magnetoencephalography or MEG. These machines do not even touch the head and look more like hair dryers (like the ones used to give grannies their 'blue rinse').

For many years I thought life was filled with coincidence, a good example is my brother from another mother who married a northern lass and moved *"op north"*. We stayed connected by driving 400 miles each way a few times a year, but time ticked, children grew, one of us is now bald and the other grey but we always stayed connected. Then after the physical visits became fewer, we found ourselves contacting each other at the strangest times. His child would be ill, but I would phone, my wife fell ill he did the same, his father passed it happened again, but it was no coincidence. Synchronicity exists but if you have doubts let us look at the numbers.

Lottery
On the 11th September 2002, many US citizens resonated their thoughts regarding the tragic events of the previous year (but forgetting building 7), they collectively had a conscious awareness. One of New York Lotteries on the morning of Wednesday 11th September 2002, a full 12 months to the calendar day of the infamy something happened selected the winning sequence as 9-1-1 (it involves choosing a three-digit sequence between 000 and 999 with the winning draw determined by numbered balls circulating in a machine). There are 1000 possibilities, not huge odds. But the fact that 5,631 had won is worth consideration or put it another way 5+6+3+1 = 15 and 1+5 = 6.

Then on the evening of 12th November 2001 the lottery number 587 was drawn in the USA, the very same day as flight 587 crashed earlier into downtown Queens, New York. Oddly, the numbers continue 12/11/2006/587, 1+2+1+1+2+6+5+8+7 = 33 and 3+3 = 6.

On Wednesday 23rd March 2016 in the UK there were no Lotto winners, but incredibly there were 4,082 players who managed to match 5 main numbers. In context just 30 ticket holders matched 5 main numbers on Wednesday 15th March 2017. The main numbers were 7, 14, 21, 35, 41 and 42, five of which are multiples of 7, which is considered a lucky number in the West. As the pay-out for the Match 5 prize tier depended on the number of winners and the amount of money in the pot, everyone picked up £15 each and you get 1+5 = 6. Had the 41 been a 28, it is possible that those 4,082 players would have shared the jackpot, presuming they had all decided to play consecutive multiples of seven but 2+8+4+8+2 = 24 and 2+4 = 6.

Another thing to consider is the *Placebo Effect*. The so-called medicine pushes pills but keeps quite about positive thinking creating a strong connection between brain and body. The pill does not fix you; the mind heals you. Medicine men (and women, *Brian – Monty Python*) will never openly reveal to you the *No-cebo Effect* (the word is even in this spell checker). Medicine knows that if you believe pills will not help you, they will not. Equally some doctors will not tell you about negative side effects of medication or procedures as they believe it is their duty to give you the best chances and occult the information from you, how often do you read the leaflet inside the prescription packet. Those of you who are deep thinkers consider, plausible deniability for a medic to claim should they administer a vaccine that harms.

Deoxyribonucleic Acid and Apes
The school of DNA is based on 23 chromosomes, but there is an interesting chromosome numbered 22 (humans normally have 2 copies of this in each cell) and the number of letters in the Hebrew alphabet is 22. Biblically the story of Noah bringing animals 2x2 into the Ark, I propose is in relation to the storage of DNA.

Psalm 139:13-16 *"For I am fearfully and wonderfully made my substance was not heed from thee, when I was made in secret, and curiously wrought in the lowest parts of the earth... Thine eyes did see my substance yet being unperfect and in thy book all my numbers were written, which in continuance were fashioned, when as yet there were none of them"*. Was David referring to DNA? I gave this some thought in terms of the biblical context which took my thoughts back to Adam giving a rib to create Eve. Look at the Deoxyribonucleic Acid, can you see it? deoxy**RIB**onucleic acid. Now consider mRNA oh messenger RNA which is a ribosome – **RIB**osome.

There is however something deeper, chromosomes in DNA are 2 pairs or 2 x 23 pairs so now you see the 223 and inversed 322 used by the mystery schools and the number of skull and bones.

The average Briton will eat many cells for nourishment, on average a human will eat 7000 creatures in a lifetime. This includes 1,126 chickens, 11 cows, 27 pigs, 80 turkeys 30 sheep, and 4,500 fish. Statistically more fish are killed for consumption than all the lamb, beef, chicken, and pork combined. Americans eat a third more. Humans kill 150 billion animals each year, should we humans carry out the culling of ourselves at the same rate of numbers then humanity would be wiped out in 17 days. Humans truly are consumers.

Chapter 13 - Masonic Ciphers and a Code to Crack

Secret Societies

The origins of these mystery schools are steeped in allegory, protectionism, and secrecy. I do not like secret groups, and you probably do not like them either which is the reason you are pouring your eyes over these pages. It is my view that secret societies were originally egalitarian but have subsequently become corrupted. Throughout the 18th century hundreds of thousands of people (predominantly men) were members of hundreds of different orders. These clandestine groups were remarkably inclusive, they welcomed nobility and merchant alike which created a danger to the establishment and the state. While originally the implicit threat to authority and the Catholic Church was demonstrated in 1738 with Pope Clement XII forbidding all Catholics to join Masonic orders. My research has led me to the opinion that these networks were infiltrated and corrupted by the very authoritarian groups that wished them to cease sharing knowledge with the everyday man/woman and were viewed as terrorist organisations with the secret manifestoes.

The difference between cryptography and gematria is simple, spies for the crown, church and indeed the Freemasons used cryptography or secret coded messages, while gematria is the number relationship with dates and words using a letter in relation to a number e.g. A=1 B=2 and C=3 etc. This is explained in the gematria chapter.

Original Templar Codes

This was known as Pigpen Cipher, mathematically known as *substitution cipher*. Pen was its original given code because pigs were kept in square shaped pens, you will see how these pens or several of them can be penned in. Let me show you how this works, you may need to read this a few times for it to 'click'.

The scaffold of the secrecy starts with 2 sets of pigpens, 1 horizontal and the other diagonal as you can see below. This gives us 13 pigpens.

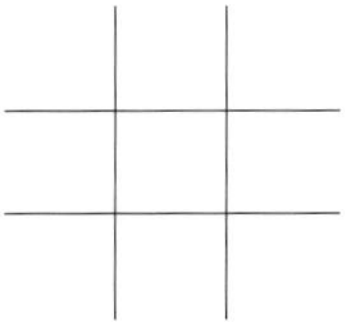

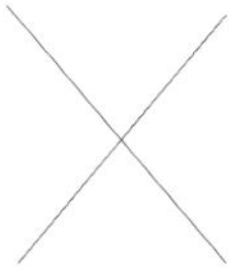

The alphabet would be placed within each pen, thereby containing all the alphabet.

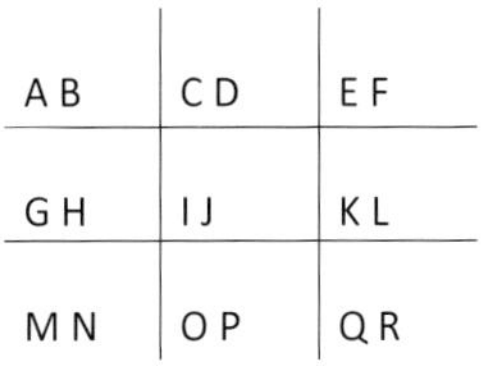

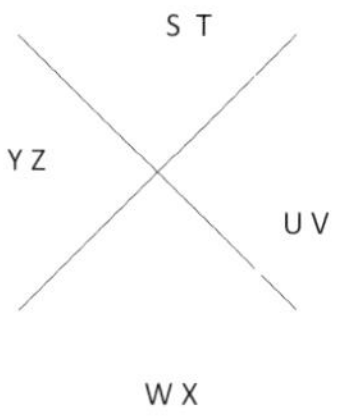

Codes or cipher is then added for half the letters, in this case the second letter in each pen.

A ±	C ⊥	E ±
G ±	I ±	K ±
M ±	O ±	Q ±

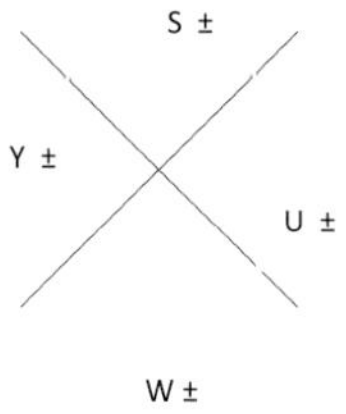

The first letter is replaced, so the entire alphabet is coded within each pen.

÷ ±	÷ ±	÷ ±
÷ ±	÷ ±	÷ ±
÷ ±	÷ ±	÷ ±

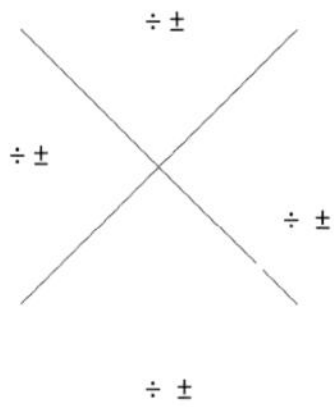

Now look at the pigpen, note the code and then you can work out the letter.

Represents A

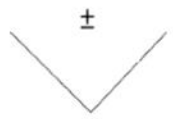

Represents T

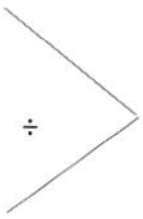

Represents Y

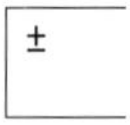

Represents L

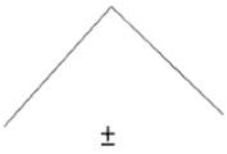

Represents X

Represent Q

A scholar would then be able to make sense of it. As you now know the basics, how about we move the alphabet along within the pens. This creates a code run.

S T	U V	W X
Y Z	A B	C D
E F	G H	I J

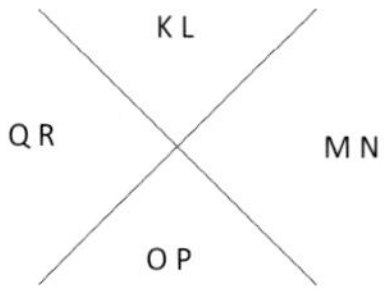

These code runs or changes where pre-arranged, happened once a month and/or on certain dates therefore cracking the codes becomes ever more difficult in the days when there was no computing power.

Alan Turing and all his work surrounding the Enigma Machine in WWII was the basis of this code cracking. Knowing the language is necessary in the first instance so you can recognise the most used letters in that language and look for the repeating patterns, for example, the most used letter in English is ‘E’. The downfall for the German military was that at the end of every coded message it ended with *Heil Hitler*, which changed daily 1 letter along. With computation Mr Turing worked out the previous patterns to then work forward, thereby cracking the future coded messages of each day.

The Knights Templar, Kings and Queens heavily used ciphers like MI5 do today in computing. If you combine magic squares and pigpen cipher in cracking code, from a letter you can see how overly complicated it can become. One of the greatest spies and the original MI5 secret service was Sir Francis Bacon. He used a cipher on the English Elizabethan alphabet of 24 letters *I* and *J* were counted as 9 and *U* and *V* as 20. With *Z* being 24. I am sure at this point your head is spinning.

The word *Sir* or *Lord* these days is a prefix of distrust and the same can be said for Sir Francis. One of his favourite numbers was 33, as his key cipher and centred around Freemasonry. The key 33 was *G*, as mentioned in *stopper 1* with the ice relationship of 33 degrees. This is where the term G-man comes from, his agents were called the *Francis Bacon Society.*

The church hated ciphers, not because the Jesuits were numerically ignorant but because they can become complex which takes time to cipher or decipher. Any form of secrecy was a challenge to the church.

Messing with Prime numbers

37	36	35	34	33	32	31
38	17	16	15	14	13	30
39	18	5	4	3	12	29
|						
40	19	6	1	2	11	28
41	20	7	8	9	10	27
42	21	22	23	24	25	26
43	44	45	46	47	48	49…

See a pattern…No?

```
37                                    31
      17                        13
            5             3           29
      19                  2     11
41          7
                  23
43                        47
```

Well now you are lost, but let me drop in the code

```
37-   36-   35-   34-   33-   32-   31
 |                                  |
38    17-   16-   15-   14-   13    30
 |     |                       |    |
39    18    5-     4-    3    12    29
 |     |    |            |     |    |
40    19    6      1-    2    11    28
 |     |    |                  |    |
41    20    7-     8-    9-   10    27
 |     |                            |
42    21-   22-   23-   24-   25-   26
 |
43-   44-   45-   46-   47-   48-   49…
```

Follow the lines and see a pattern

```
37     -     -     -     -     -    31
|                                   |
|     17     -     -     -    13    |
|      |                       |    |
|      |     5     -     3          29
|      |     |           |     |    |
|     19     |     0     2    11    |
41     |     7     -     -     |    |
|      |                            |
|      |     -    23     -     -    -
|
43     -     -     -    47…
```

These are the atoms of the math world.

For the real mathematics hounds, below is a quote from Shakespeare using the English Ordinal Gematria, see if you can crack the code:

20,1,11,5 20,8,5 6,9,18,19,20 12,5,20,20,5,18 15,6 5,1,3,8 19,5,20 15,6 6,15,21,18

3,14,2,5 12,20,8,5 15,1,18,5 20,1,19,8 8,3,2,24 5,26,13,13 19,24,26,21

13,1,2,5 1,13,1,16 11,10,5,2 5,16,23,13 20,1,23,5 8,13,3,21

20,11,5,14 8,23,12,15 5,15,23,2

13,15,5,14 1,23,14,20 14,17,1,23

7,8,15,19 15,5,7,9

20,2,19,20 15,18,14,3

16,17,3,5 1,6,12,20 7,11,15,9 5,21,13,17

15,22,11,26, 14,11,22,8 5,12,21,11

26,3,4,20 5,16,18,4 18,8,2,19 15,4,16,19

15,20,17,9 14,24,26,7 5,24,22,11

Chapter 14 - Hate Numbers, Hanging and Jeans

Society increasingly uses the saying *"hate speech"* and equally governments try to clamp down on its very existence. This in turn has served to force 'hate' groups to communicate in secret by using numbers. The coding that is used is increasing year on year as they are taking a leaf out of codes used by many ancient mystery schools and secret societies. Often these numbers are demonstrated on clothing, bandannas, or tattoos.

Turn to the gematria codes at the back of the book using the Simple English Gematria or English Ordinal Key. These are the codes that drive some of these numbers.

12
The *Aryan Brotherhood* in the USA use this number. The 1 stands for *A* and the 2 for *B* hence the term Aryan Brotherhood.

13
Now being used by the *Aryan Circle*, a large group from Texas emanating from a prison gang. The number 1 and the number 3 equals *A* and *C* in the Simple English Gematria, therefore representing the first letter of each name.

13/52 and 13/90
Used in Germany and the USA these numbers are used in hate form against African Americans. White supremacists claimed that people of colour make up only 13% of the US population and commit 52% of all murders and 90% of all violent are interracial crime. All the statistics are arguable, but they use these numbers to drive their codes.

14
This one is a little more complicated as it stands for the white supremacist slogan of '*we must secure the existence of our people and a future for white children*'. The 14 is numerical shorthand for how many words is used in the statement.

The term '14 words' is a reference to the above quote and appears on some T-shirts.

14/23 or 1423
Used by the *Southern Brotherhood* the largest white supremacist prison gang in Alabama, the number 14 is the reference to the '14-word' slogan as above. The 23 refers to the 23 precepts, a list of rules that they must follow.

14/88 or 1488
Used by many groups we have the typical '14-word' statement but this time the 88 stands for *HH* which of course is Heil Hitler.

18
This is another numeric code for Adolf Hitler, 1 equals *A* and 8 equals *H*. This number is used mostly in the UK by *Combat 18,* they add the *C* for Combat and you get C18.

21-2-12

A Floridian based prison gang called *The Unforgiven* use these numbers; 21 being *U* for *Unity,* 2 for *B* in *Brotherhood* and 12 for *L* meaning *Loyalty.*

23/16

Used on the West Coast of America this equates to 23=W and 16=P which stands for *'white power'*.

28

Now that you are starting to understand the numerology, this one is quite simply *blood* and *honour.*

38

Reputedly created by a group called the *Hammer Skins* they are another USA skinhead group. The letters stand for *CH* or *Crossed Hammers* which is the groups logo.

43

This one was created by the *Supreme White Alliance* which in turn stands for *SWA* or *19,23,1* and when you add these together you get 43.

83

H and *C* which stand for *'Hail Christ'*.

88

HH or *'Hcil Hitler'*.

311

11 meaning *Klan* and 3x11 is *KKK.*

318

3 meaning *C* for *Combat* and the 18 identifies *Adolf Hitler,* it represents *Combat 18* a British white supremacist group.

511

Gang of the *European Kindred,* obviously the two drivers here are *E* and *K.*

737

This group are *Public Enemy Number 1* based in California and a white supremacist group. This had me a little confused until I looked at the letters *P, D* and *S* on an old mobile telephone keypad and realised it was for *Peni Death Squad* which is another name for the group.

1331 or 336

The 11th letter of the alphabet is *K,* this one is quite simply *Klu Klux Klan* (Greek – *Kyklos* which means *"circle"*) and the Scottish-Gaelic word "clan". 11x11x11=1331. Sometimes used as 33-6 with the 6 meaning the historical era that clan bases its history on.

5 words / 5W

This turn of phrase is used for advocacy against the police when being questioned, it simply means *"I have nothing to say"*.

Jeans and 10

Within popular urban culture young men sometime wear baggy jeans worn low, about a foot down from the waist and the term for this is SAGGING or SAGGIN jeans. Whilst most correctional facilities forbid the use of belts, the origin of this fashion comes from the mid 1970s from a prison in Joliet Illinois where 2 homosexual men started the trend of demonstrating to other prisoners who was available for anal penetration.

Research further and you will find the name Willie Lynch a British slave owner in the British West Indies who gave a speech about colonial slave control in Virginia in 1712. He had 10 rules for keeping a slave. Ranging from 1 to 10, these fixed rules ascended with a level of cruelty and were adopted by many plantation owners. Rule 10 was to rid the plantation of one and to demonstrate to the others 'The Rules', it involved hanging. This is where we get the term *Lynch,* via Lynchburg (Virginia, USA) named after his slave quarters and his family.

It follows that in law a free man is *hung* but a slave is *lynched*.

Another was Rule 9, to strip the slave naked over a tree stump and rape them; called *'breaking the buck'*. For months after this treatment the slave would have to walk around with sagging britches to demonstrate they had been punished. Things are always 'In Plain Sight' you just must seek them out. If you spell SAGGIN backwards it is a terrible word that is sadly not lost on me.

Virgin Islands

In the year 1493 Christopher Columbus while searching for sail routes to India sailed into the currently known United States Virgin Islands (USVI) and called them the beautiful Virgins. This, however, is not lost in history in his reasoning. The legend of the martyred St Ursula and her 11000, 1100 or 11 virgins has many accounts to her origin. Some are that she was a gift to join kingdoms in her marriage, others that she was a child lost at sea. Columbus named the islands after this legend and the many smaller islands attached in this area representing her accompanying Virgins.

The Danish West India company took possession of the islands in 1666 beating Holland, France, India, Spain, and the Knights of Malta all of which wanted to claim the islands. Its use of slaves made the islands of intrinsic value in the production of cotton, sugar cane and rum. The abhorrent value placed on a slave throughout the Caribbean and America was set at £729 the value of a male slave at the time of the "Slavery Abolition Act of 1833" in today's money was £360,000. Slaves and their revolts, genocide, poverty, piracy, rum have deep seated histories with this island chain including some present-day occupiers like the late Epstein. Richard Branson owns Necker Island (Latin-Neca to *put to death/kill/destroy*) in the British Virgin Islands and has his empire is called Virgin. These island chains have a sinister history.

Chapter 15 – Shoe Laces, DNA and Death

The human genome centres around DNA strands which are helical strings manifesting in a vertical pattern. It is a mathematical setup of continuous loops. DNA shoelaces form a two-line segment with a chosen path running through eyelets. The lacing of a shoe has a direct relationship with DNA formulation in its pattern which is called the *pollster*, in its general description it is how tight you can pull 2 opposing eyelets together. The telecommunications industry also uses this method, but it is called the *Combinatorics* which involves calculus graph theory and geometry. The mathematics are in depth and overwhelming, if you have 8 pairs of eyelets you then have a total of 52,733,721,600 choices on how you may want to lace them together.

These patterns have many names and here are a few:

- Criss Cross, Over Under, Gap, Straight European, Straight Bar, Straight Easy, Hiking/ Biking, Quick Tight, Gipo, Ukrainian, Corset, Sawtooth, Lightning, Shoe Shop, Display, Chevron, Ladder, Spider Web, Double Back, Bow Tie, Army, Train Track, Left Right, Double Helix, Double Cross, Hash, Lattice, Zipper

Ties or tie knots also have mathematical structure behind them. Usually, children learn how to tie a tie at school and in my experience once a knot formation is tutored, people stick to that the rest of their lives. The most used is the *Oriental knot,* mathematically formed by LoRiCoT. It is the code formation which people store in their mind as a pattern when they tie the knot, but it means *"Left Out Right in Centre Out and Tie"*. The four-in-hand-knot is LiRoLiCoT or *"Left in, Right Out, Left In, Centre Out and Tie*". I was taught the *Windsor knot* and naturally carried out the LiRoCiLoRiCo and Tie not knowing .

DNA

DNA is an extremely long double-stranded rope or lacing in which the two strands are wound about one another. As a result, topological properties of the genetic material, including DNA underwinding and overwinding, knotting, and tangling, profoundly influence virtually every major nucleic acid process. The DNA in your cells is packaged into 46 chromosomes in the nucleus.

The 3 key mathematical concepts: twist (Tw), writhe (Wr), and linking number (Lk). Twist represents the total number of double helical turns in each segment of DNA.

Etymologically, the word *'spiral'* springs from ancient roots inextricably bound up with ideas of creation, life-giving and aspiration. From the Latin *spiralis* or *spira,* and the Greek *speira*, meaning a spire or coil, a conical, pyramidal structure, as well as from the Latin *spirare*, meaning to breathe.

The Noose

Commonly known as the *hangman's knot,* the occult number that hides behind is 13 and extensive mathematics goes into calculating these killing devices. A *slipknot* is designed to tighten when more weight is applied to it in a vertical position. While you would recognise a noose, have you ever considered the number of coils that hold the noose together? Surviving a hanging (lynching mentioned elsewhere in this book) resulted in you being set free, as it is considered an act of God in intervention.

Much thought is placed into the origins of a noose which centres around the number of coils that hold the noose. Depending upon the thickness of the rope and the drop of the gallows comes into this calculation. If the drop is too long it results in decapitation, if it is too short it results in strangulation. The science behind the coils is to push the head to one side bending the vertebrae causing it to snap the neck and dispatching the individual quickly and preventing God's intervention. The ideal number of coils is 8 for a quick death, 13 will result in slow strangulation. This was invented by Harrold Gundry of Bridport, England in the late 18th century.

The *Official Table of Drops* created in 1886 by the British Home Office for the discovering and reporting on the effective manner of hanging is used around the world today as the perfect chart of dispatch. The perfect foot pounds of drop in energy terms are 1,260.

In the song by Woody Guthrie the lyrics are as follows "*did you ever see a hangman tie a hangknot? I've seen it many a time and he winds, he winds, after thirteen times he's got a hangknot*".

The next time you tie your shoes the matrix numbers are right in front of you.

Chapter 16 - Movies, Easter Eggs and Milk

All storytelling is based on the number 6, 3 and 2

Words have emotion (books) as does viewing something (pictures) and combined you can create an experience. Humanity has always passed on knowledge, facts, and events by storytelling. This started around campfires which became great halls of kings, then taverns to travelling shows, music halls, cinemas, IMAX screens and now immersive video gaming. Humans like to be transported with their imagination.

I introduce to you *Sentiment Analysis*; it is based upon mathematical emotional arcs or graphs in any tale that is told. All storytelling is based on 6 options and within these there is either a pattern of 2 or 3, let me explain:

1. Fall/rise/fall – pattern of 3, the movies that have this are:
   District 9, Interview with a Vampire, Titanic

2. Fall/rise – pattern of 2, failure to success:
   Constantine, Limitless, Silence of the Lambs, Groundhog Day, Shawshank Redemption

3. Rise/fall – pattern of 2, success to failure:
   John Wick, Apocalypse Now, The Founder, The Laundromat, The Wolf of Wall Street, The Theory of Everything, Jurassic Park

4. Steady/rise – pattern of 2, rags to riches:
   Trading Places, King Arthur, My Fair Lady, Pretty Woman, The Golden Fleece, Raiders of the Lost Ark

5. Steady/fall – pattern of 2, story of tragedy:
   Romeo and Juliet, Seven Pounds, Dead Poets Society, Legends of the Fall, Last of the Mohicans, Dances with Wolves, Forrest Gump, Marley and Me, Ghost, Carol, Green mile, Frankenstein, Dracula, Thelma and Louise

6. Rise/fall/rise – pattern of 3, emotional riches, emotional fall, emotional rise. Boy meets girl, boy losses girl, boy wins girl:
   Cinderella, A Christmas Carol

The most popular and profitable movies are either - fall/rise/fall or rise/fall arcs. There are some who believe there is a 7th which covers comedy as its own complete idea, but within these movies the plot can relate to the *Sentiment Analysis* of the 6.

Religious Books

Movies based on books with malevolent intention are included in many films such as:

- *Blaire Witch 2, The Evil Dead, The Mummy, Hocus Pocus, Harry Potter and the Philosopher’s Stone, Harry Potter and the Prisoner of Azkaban, Pan’s Labyrinth, The NeverEnding Story, Ella Enchanted, Dungeons and Dragons, The Colour of Magic,*

Maleficent, Residue, On the Count of Zero, Shang-Chi- Legend of the Rings, Practical Magic, The Pagemaster, Inkheart

There are, however, 3 books never considered as occult and these are the Quran, the Torah, and the Bible. WHAT? I hear you cry! Consider that these books balance stories of good and evil, they have incantations as prayers, they offer secrets to be found. Let us not forget occult means hidden.

Underlying Themes

Movies have underlying storylines for example *Spider Man* - growing up, *X-Men* - the struggle for Civil Rights, *Lord of the Rings* - the creation of a totalitarian regime, *The Matrix* - a conspiracy theory, *Toy Story 3* - a Holocaust and *Groundhog Day* - the rebirth, belief of Buddhism. The esoteric cannot be separated here.

In the movie *Sunshine*, the space conveyance was called Icarus. Obvious in terms of the legend of a man who affixed feathers and molten wax to form wings, that took him too close to the Sun which then melted and ended in his demise.

Jurassic Park is based on the creation of dinosaurs from the original DNA number code but within chaos theory mathematics. The scientists decide not to use male DNA but only female thereby trying to control breeding at the dinosaur park. The basic mistake was to use a cell line of a frog which can be androgynous (male and female). The message throughout the movie was that 'nature will always find a way to reproduce'. The allegory message was obvious to me when Professor Alan Grant (Sam Neill) tries to buckle himself into his seat during turbulence on the helicopter ride to the island. He can, however, find only 2 female buckles on his seat belt. Not being able to find the male he simply ties the ends of the 2 females together therefore finding a natural solution to securing himself in. He found a way to make it happen. This clear allegory was the entire basis of the movie where the mathematics of nature will always find the solution.

Cloverfield ended with the characters being buried under a New York bridge and if you listen to the end of the credits there is a garbled message. This irritated me until I carried out backmasking (playing it backwards) as visited in *stopper 1* and the voice is clear *"It is still alive"*. In the movie industry these things are called *'easter eggs'*. Fun, right? I would argue NO for sometimes they are laid out for people to crack codes and some of these can be sinister.

We are going to break the first and second rule of *Fight Club* because we are going to talk about the *Fight Club.* To gain admission to the club's temple house, candidates had wait outside for 3 days without food, encouragement, or shelter. Being berated is a test but in esoteric terms the 3 days represent the 3 first initial levels of Freemasonry. In the mystery schools the member who guards the temple outside is called the 'Tyler', hence, the name of the character Tyler Durden. The secrecy to Masons is to not discuss their internal affairs outside the temple hence the 1st and 2nd rule of fight club. The female in the movie is called Marla Singer, a name that implies she cannot be trusted and will *"sing like a canary"* if she finds out any secrets. The movie is about the destruction of the existing control system and the birth of a new one, using a secret society of men. It is also about a huge underground movement of men called MAGTOW – *Men Going Their Own Way* and females more often than males do not even know of its existence. It is an anti-feminist movement for males who no longer want anything to do with women but only sex and that only on their terms. From the mistreatment towards men in family courts, to male suicide rates, the movement is growing fast. Not stopping for women broken down on the motorway in case of accusations or opening doors for them in case of being called a misogynist, this is spawning a huge underground male movement. "We do not talk about fight club" now means something completely different for females reading this book.

International Business Machines - IBM is a nice fluffy corporation but if you want to know their origin look up the Hollerith Machine and the Nazis for it just may change your mind. In Stanley Kubrick's *2001: A Space Odyssey* the computer goes psycho and tries to kill off the crew. In the Pan Am Clipper scene, the IBM logo is at the centre of the cockpit instrument panel. The computer's name is HAL (Heuristically Programmed ALgorithmic Computer). Look at the alphabet and you will see that HAL is IBM followed by its predecessor letter e.g., H is followed by I, A is followed by B and L is followed by M, so IBM and HAL are linked. While IBM through its German subsidiary and their computers provided help to the Nazis in mass genocide and HAL goes rogue and tries to kill humans itself. In this tale the astronaut shuts down the computers data cards with a screwdriver and as HAL starts to fail, he sings "*Daisy Bell*" which was the first song chimed by IBM's synthetic speech computer (IBM 704), so these hidden messages run deep.

The *Truman show* is based on control of an individual and the monitoring by the authorities. In the scene where Truman and his father are lost at sea, the ring his dad is wearing is the all-important part of the movie. There is no reference to his father handing him the ring before his staged demise, but Truman ends up wearing it throughout the movie as a homage to his father's memory. This ring had a camera and a verbal monitoring device which in turn allowed the controllers to have the constant surveillance on him. It is only later in the show that Truman gives back the ring to his father who is caught on set and thereby removes his monitoring device and gains his first step towards freedom.

Pulp Fiction is the most religious movie you could ever watch. The central theme is the search for a mysterious briefcase for Marcellus Wallace, a cartel gangster. His office is run out of the topless bar named Sally LeRoy's, in French *Le Roy* means 'The King'. His 2 soldiers (Knights) are Vincent Vega and Jules Winnfield who are on the quest for the briefcase, the contents of which is never seen. Now consider that the combination to the briefcase is 666 and its contents are always glowing golden on the faces of those that peer into it. In Revelation the mark on the Antichrist is 666 and is this the reason Marcellus has a plaster on the back of his neck?

The wife of Marcellus is his property, called Mia – Italian - *mine* and Vincent fears this fact for in his violent world this is the equivalent of possession or being possessed. The religious connotations continue for just before Vincent dispatches the young men in the opening scene, he quotes Ezekiel 25:17 "*The path of the righteous man is beset on all sides by the inequities of the selfish and the tyranny of evil men. Blessed is he who, in the name of charity and good will shepherds of the weak through the valley of darkness, for he is truly his brother's keeper and the finder of lost children. And I will strike down upon thee with great vengeance and furious anger, those who attempt to poison and destroy my brothers, and you will know my name is the Lord, when I lay my vengeance upon thee".*

The subsequent reign of bullets has flashing in the film as their souls are taken, which is the same golden colour you see within the briefcase. Therefore, if the case does contain a soul whose soul could it be? Is Marcellus trying to take his back from the devil? In the pawn shop sequence Butch makes a run for it and drives into the Mason and Dixon gun shop, the relevance is that the Mason and Dixon line was the dividing line between the slave and the free states in the American Civil War. A direct reference to freeing men from bondage.

In the final scene Jules states that he is going to stop doing *"the dirty work"* thereby essentially turning his back on the devil. Vincent, however, disagrees and continues and due to the back tracking of the linear story line he dies as a result. The movie is about religious redemption for Vincent and Lance save Mia, Butch saves Marcellus, and the wolf saves Vincent and Jules. The ultimate 'angel' however is Jules who saves himself, turns his back on the devil and 'walks the earth' saving others.

Paul is another religious movie that you may not have realised. It is based on a sci-fi road trip with an escaped alien, filled with comedy elements such as being anally probed and the little man likes to smoke weed. The alien converts Kristen Wiig's character from a Christian fundamentalist into a nonbeliever. Her T-shirt has the logo 'Evolve This' and it depicts Jesus shooting Darwin with a gun. Before Wiig begins to accept Paul as an alien, she calls him a demon. She is convinced that aliens do not exist but due to her religious background demons do. Paul like Jesus has healing powers for he resurrects a dead bird (then eats it), heals Wiig's long deformed eye (giving site to the blind) and finally saves Graham Willy's life which nearly kills him, leaving the inference 'dying to save others' in sacrifice and charity. The apostle 'Paul' is a character from Christianity who went through dramatic conversion on the road to Damascus where he met God and became a believer.

Alien is a feminist movie genre about foetal termination and rape. To impregnate humans the face hugger inserts its proboscis down their throat forcing the victim to be a host for the unborn. Many women throughout the centuries have died in childbirth. When the alien is born the males die and suffer the same that millions of women have done before him. At the end of each movie the female always out lives the males through courage and intellect.

Easter Eggs

Film directors love hiding these *'easter eggs',* let us start with Stanley Kubrick who directed several movies and was a huge occultist in his cinematic triumphs. Richard Nixon appointed Donald Rumsfeld as director of the US Office of Economic Opportunity and was said to have considered *Dr Strangelove* as one of his favourites which directed by Stanley Kubrick. Donald knew Kubrick who was just finishing the filming *2001: A Space Odyssey* at the MGM studios in Borehamwood, England.

According to Pentagon records Nixon was determined to film the Moon landings but the technology was not available, therefore, a meeting was arranged with Stanley Kubrick. Rumsfeld, Kissinger, and Nixon met with Kubrick to explore what would be required to film the event. A few weeks later the CIA demanded access to the studio Kubrick was using for the

2001 movie to dry run some of the filming required on the Moon, at first Kubrick refused only to change his mind a week later. Now consider that Kubrick filmed several genres up to this point including a war movie *Paths of Glory,* a comedy *Lolita*, *Spartacus* a rebellion epic and the political thriller *Dr Strangelove* and was about to finish a science fiction movie *2001: A Space Odyssey* but had the CIA on his set. The question I have is did something happen and did he need to leave an easter egg for everybody to see. Of all the other genres of movies he had filmed he had never visited horror.

In 1980 *The Shining* was released, and I noticed a number. In the corridor scene, the child Danny (actor Daniel Edward Sidney Lloyd) starts off by playing with toy cars on the floor, and when he then sits back you notice on this home-made woollen jumper a rocket which reads Apollo. Nothing much to see here until he stands, and it says Apollo 11. When he stands it is like a rocket leaving the ground, then it cuts to him walking down an exceptionally long corridor, but he does not make it to the end of the corridor.

The Apollo 11 mission was launched on 16th July 1969 from Cape Kennedy as the first crewed mission to the Moon with Buzz Aldrin, Michael Collins, and Neil Armstrong. If we go back to the corridor scene above and the child walking with the rocket on his jumper, it depicts a passage of the rocket's trajectory. Halfway down the corridor, the child sees an open door to his right, and it has a key with the number 237. In the novel the room number was 217. Danny does not complete his mission down the corridor.

Why would a director place occult images into a movie and make it obvious by the number difference on the door? What need was there to change the number? What was the director indicating?

The Moon is a natural satellite to the Earth, with an elliptical orbit averaging circa 381,600 km – 382,900 KM. If we convert this to miles, it is 237,000 so does room 237 represents the Moon. Apollo 11 (the boy) it seems enters room 237 but it is not at the end of the corridor and the full distance has been limited in full travel. Interestingly the Cape Kennedy launch was at 13.32 UTC on 16th July 1969. After 2 hours and 33 minutes the lunar module detached for the moon mission (13.32 33 mins). So, we see 233 and 3333.

In 2018 *Searching* was released, a tech thriller following David Kim who becomes a cyber detective while searching for his missing daughter, appears straightforward does not? If you have not seen this movie, watch it as there is a subplot going on and most people never notice. In every scene there is a reference to an alien invasion going on around the world from computer screens, newspapers, and the media. It truly is a skilled occult.

In Ridley Scott film *Alien,* 2 moons played an enormous role being LV-223 and LV-426. The former jumped out as it is reversed 322. In the 1979 movie, the crew of the Nostromo visited LV-426 and in 2012 the movie *Prometheus* involved the crew landing on LV223. The latter is the planet where the engineers created all the problems with the creation of the *'Xenomorph'*.

In *Prometheus,* the planet LV223 seems to point towards the 3rd book of the Bible, Leviticus 22:2-33 the Lord said to Moses "*Tell Aaron and his sons the gifts that the Israelites bring to me become wholly. They belong to me, so you priests must show respect for these things. If* *you do not you will show that you do not respect my holy name. I am the Lord if anyone of your descendants touches these things that person will become unclean. That person must be separated from me. The Israelites gave these things to me I am the Lord Aaron was Moses right-hand man and he has his sons were gods first priests ministering to the Israelites god's chosen peopl*e". It appears that the engineers were *'Nephilim'* who genetically engineered humans and they in-tern created the artificial persons or robots.

In the *stopper 1* I wrote about the esoteric messages in Disney movies with sex but what about numbers, so here it is once more with examples that the cartoonists really layered into the children's entertainment with sinister messages numerically.

Alice in Wonderland fell down the rabbit hole, the grandfather clock she passes in her decent has its hands at 3 o'clock. In esoteric terms this is the witching hour or the time at night when the power of darkness is said to be at its most powerful.

Then there is A113 in *The Brave Little Toaster,* the apartment number where "Master" lives; *The Princess and the Frog,* a trolley is labelled with the number; *Lilo & Stitch*, car registration number; *Toy Story*, car number plate on Andy's mom's cars; *Monsters University*, number on a door; *The Simpsons*, various but one is Bart's photo crime number; *The Iron Giant,* damaged truck number plate. Is this something sinister, no, it is an easter egg in media and developed by alumni of the California Institute of the Arts. It refers to a classroom when the animators were taught.

In the *Ray Donovan* series, actor Frank Theodore Levine (Ted) plays a criminal called "Bill Primm" a casino owner who is also a serial killer with a reputation for throwing his victims into wells. Looking deeper (no pun intended) Ted Levine played a serial killer in *Silence of the Lambs* who kept his victims in, well in his basement for a forced weight loss program so he could loosen and shrink their skin to later fashion clothing from the epidermis. In the movie the FBI gave him the pseudonym "Buffalo Bill". In the *Ray Donovan* series, his name was Little Bill who lives in Primm (season 4). Going deeper the events take place around a town called Primm in the State of Nevada USA where a real casino exists called Buffalo Bill's Resort & Casino.

Tenet - 911

Do you remember the SATOR square? Then if you watched the movie "*Tenet*" you may now realise the depth to which some directors work.

S	A	T	O	R
A	R	E	P	O
T	E	N	E	T
O	P	E	R	A
R	O	T	A	S

TENET is the name of the protagonist, SATOR is the name of the villain and the security company he hires is called ROTAS. The introduction of the movie takes place in an OPERA house and the final word in the movie is AREPO. The entire movie is about the inversion of time, if you watch try and see if you notice the 5 pentagrams that make up the layout of Freeport. These freeports are a place where art work is transacted in secret like off shore banking and governments cannot control them. In the movie they are looking for something in the centre of the freeport. As a teaser look up Revelation 9:11 and you will see where this led me. Then consider the fact that on 11th September 1941 construction of the pentagon building started with the attacks being exactly 60 years later, to that day. This movie takes on a whole different meaning.

Bond Movies
In this 1959 movie about 007 and the protagonist *Goldfinger* is obsessed with his surname first 4 letters. The hidden meaning here refers to King Midas who has the golden touch, hence the use in his surname the word *finger*. In the movie his first name is Auric from the route Latin - *aurum* meaning 'glow of sunrise' or 'gold' with the alchemical symbol for gold being Au in the periodic table. His Rolls-Royce Phantom III has the registration number AU1. In the movie 'Operation Grand Slam' was aimed at wiping out the value of Fort Knox containing most of the West's gold supply, thereby restricting the market and increasing gold's value. With the operation taking place on Sunday, for the day of the Sun and in alchemy this golden yellow process is known as CITrINITaS.

Other esoteric principles are demonstrated in the Bond movies by Ian Fleming such as *Dr. No* 1962. In a scene on the beach where the boat captain called George fights a mechanised tank with flame thrower is a clear indication of George fighting the fire breathing Dragon. This beach area is also known as Crab Key a direct indication of the Tropic of Cancer.

The overtly sexual names of females within the movies are:

- Pussy Galore - *Goldfinger*
- Mary Goodnight - *The Man with the Golden Gun*
- Honey Ryder - *Dr No*
- Dominetta Vitali - *Thunderball*
- Tiffany Case (another word for chastity belt) - *Diamonds Are Forever*
- Plenty O'Toole - *Diamonds Are Forever*
- Octopussy - *The Man with the Golden Gun*
- Kissy Suzuki - *You Only Live Twice*
- Dr Holly Goodhead - *Moonraker*

M is the 13th letter of the alphabet and in the mystery schools it represents leadership ruling over the 12 houses of the zodiac. Q is the 17th letter in the alphabet 1+7 = 8, the number of eternity and supply. Hence why he is a quartermaster and gives Bond all his gadgets. Numerological symbolism is clear if you read the books rather than watch the films, in the original novel *You Only Live Twice* Bond is renamed 7777 after the death of his wife Tracey. In the plot he is seeking information out of Tiger Tanaka and the secret cypher machine referred to as *Magic 44*. The 7th card in the tarot of the major Arcana is of war and victory over the enemy, the embodiment of Bond movies.

DeLorean
Some occults are wrong. In *Back to The Future* the famous vehicle had to hit 1.21 Gigawatts of power to time travel, this meant accelerating to 88 MPH. After my hours of number

crunching, I realised why. In 1979 American vehicle regulations had all speedometers limited to a maximum of 85 MPH and that is what showed on the display. In the movie, the speedometer is not shown, it is displayed as a digitised box and therefore the 88 can be reached visually. The DeLorean car had a length of 4,216 m, if the wormhole remained open for .1072 s, then divided the length by 107.2 you get 39.88 m/s or 88. The time travel takes place in California which is 37° North in latitude. Travelling East from 37° North and circling the entire distance of the globe would be 32005 km, the speed of light is 300,005 km/ps. The 1.21 GW of power irritated me until I realised that a Watt is a Joule per second therefore 1.21 GJ equals 1210 megajoules. If we calculate that *g* is the gravitational constant and *m* is the mass of the Earth and the Earth's diameter the vehicle would only need to reach 77 mph. It is a shame they got this calculation wrong.

Timeline

Here are some more movies that are set in the future, with most heading to a dystopian nightmare. Working through the years in which these movies were portrayed in descending order, pay attention to the year in which you are reading this book.

Movie Title	**Year movie was set**
Dune	10191
Idiocracy	2505
Zardoz	2293
Logan's Run	2274
The Fifth Element	2263
Forbidden Planet	2220
The Matrix	2199
Sleeper	2173
The Black Hole	2130
Alien	2122
Artificial Intelligence	2101
Total Recall	2084
Minority Report	2054
Mad Max Fury Road	2050
Event horizon	2047
12 Monkeys	2035
Demolition Man	2032
Planet of the Apes	2029
RoboCop	2028
V for Vendetta	2027
Children of Men	2027
Soylent Green	2022 This Year
I Am Legend	2021 Last Year
Cell	2020
Blade Runner	2019
Roller Ball	2018
The Running Man	2017
Back to the Future Part II	2015
The Postman	2013

Freejack	2009
2001: A Space Odyssey	2001
Escape from New York	1997
A Clockwork Orange	1980

The film industry has worked towards a dystopian future with this year (2021) being *I Am Legend* when a manufactured vaccine turns the world's population into zombies and according to this list, next year we will be eating each other following the script of *Soylent Green*. The previous year would have been based on the movie *Cell* with the 5G / mobile phone plot.

Milk

As food it is obviously beneficial to the young mammals. There is, however, a malevolent use of milk within movies. If a character drinks hard liquor (like whiskey, bourbon, vodka) they come across as tough individuals. James Bond drinks martinis so is seen as cool, if you drink wine, you are sophisticated and champagne well then you have money. Adults drinking milk, however, sends a different occult message to the moviegoer. Milk depicts a child or human with innocence, so the first movie that springs to mind is *Catch Me If You Can* where the hero/villain travels by Continental Airways as an adult but is really and adolescent in the same way it was used within *Forrest Gump*. James Dean was always depicted as a rebellious adolescent who drunk milk. In *A Clockwork Orange* the gang of aggressive teenagers also only consumed milk (the Adrenochrome drug will be visited elsewhere in this book).

In the professional hitman movie *Leon,* he was depicted as drinking milk, but this was a warning to the viewer that the man drinking milk should be watched carefully. Other films where this is depicted is in *No Country for Old Men* (the villain), Hitchcock movie *Suspicion* (the poisoner), *Inglourious Basterds* (Nazi SS Colonel Hans Landa which made him look creepy), *Mad Max Fury Road* (the elite had wet nurses expressing milk for the Royal Family who controlled water) and in *Meet the Fockers* (Mr Sandler drank breastmilk from the fridge, and I noticed that people were repulsed). Subliminally it is trying to tell you that when milk is sour it has gone off and when a grown man drinks milk something is off too. Other cinema extravaganzas that have you questioning the characters after seeing them drink milk include *The A-Team* (B.A. Baracus), *Lock, Stock and Two Smoking Barrels* (Big Chris), *There Will Be Blood*, and in the *Aviator* where the main character is viewed as unstable and by drinking milk it endorses the message.

4 Colours

The *Matrix Reloaded* had all the vehicle registration numbers as Bible passages for example 'IS 5416', Isiah 54:16 " *Behold, I have created the smith that bloweth the coals in the fire, and that bringeth forth an instrument for his work; and I have the waster to destroy"*. Another registration number in the movie is 'DA203' from Daniel 2:3 "H*e said to them, 'I have had a dream that troubles me, and I want to know what it means'"*. In the movies they had a hovercraft called the Nebuchadnezzar, who was the name of a Babylonian King who quoted the Daniel 2:3 text.

In the movie, Neo the only human who can see the code, the code is known as Matrix digital rain or green rain. Why green?

The computer screens are green on black as is the numbering on Neo's alarm and his telephone. When Morpheus is in the Matrix, he wears a green necktie but in the real world he does not. The wallpaper in the Oracle's apartment is green, as are all the walls in the building. The oracles outfit is green, all exit signs in the city are green not the standard red. Whenever you think of the Matrix you automatically think of green on black, it is a subliminal message of colour recognition same as the *Schindler's List* with black and white and a flash of red.

Now let us explore how the way advertising runs deep in movies.

27x41 inches is the standard size of a movie poster which contain subliminal messages.

Kong skull Island was released in 2017, *Apocalypse Now* released in 1979 and as you can see some posters call on previous successful images. Both designs replicate each other from the Sun with the helicopters in front of it, main character positioning and even the writing have direct relationships.

Occult runs deep within some posters; you must firstly notice them. As you can see in the above *The Silence of the Lambs* poster there is a skull that has been placed within the pattern of the moth. If you understand MK-Ultra and the moth's positioning over the mouth, then you are already deeper than many. Those of you who see the skull on the moth should be pleased with yourselves. Occultism, however, is a skill of seeing layers and I am about to show you another deeper meaning on this insect. Concentrate on the skull and take your time.
Do you see an image of women created by Salvador Dalí?

When you see an occult, spend more time looking at it. Within the poster we have in the title with the word 'silence' which is a reference to using MK-Ultra (mind control CIA program which started in the 50s). The 1951 work of Dalí *The Morphing Body* was about building a face out of humans and it has a direct correlation to the theme of the movie with Buffalo Bill's attempt to metamorphosis into a woman wearing the skin of his female victims.

I bet you have even seen this used on another movie poster but have not realised it.

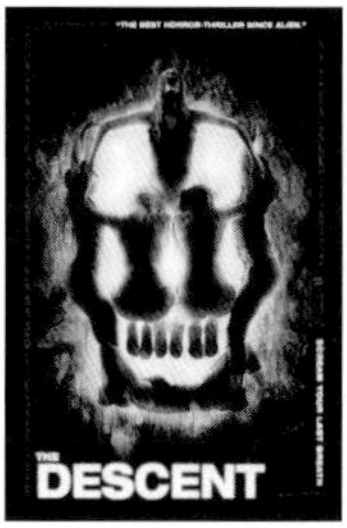

Here are some themes for you:

- Back-to-back montages are about relationships

Boredpanda.com

- Benches are about lonely people and contemplation

- Backs are shown for revenge and justice

- Face to face posters is about conflict

- Legs are about titillation or sex

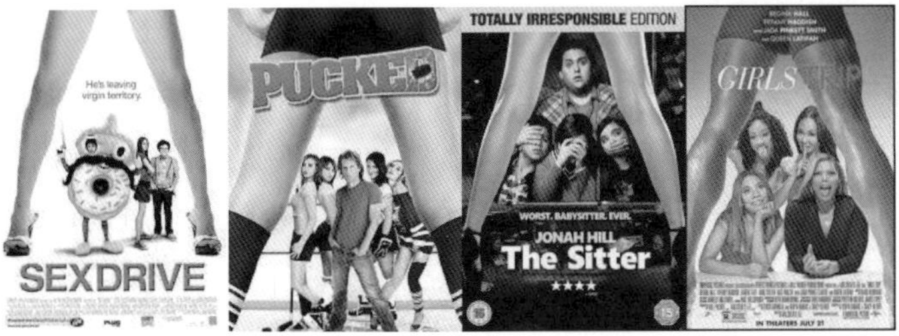

- Messages such as justice is blind can be delivered by the poster

Colours that are opposite create confusion in the human mind, like night and day, black and white, fire and water, it is based on conflict.

- Action movies can be orange and blue as follows

Subliminal signalling has more power over the human mind that most people give credence to. An example of the most complex and powerful ideas is the *Wizard of Oz* by L Frank Baum, a practicing theosophist and occultist. Colours in this childhood classic has deep metaphysical meaning with the plot being that a girl called Dorothy living on a farm in Kansas, gets swept away by a tornado and travels a yellow brick road with her dog (spirit follower) called Toto. The yellow Brick Road is a golden (money backed by gold) path seeking the Emerald City (green, the colour of the greenbacks or dollar, fiat currency) which starts with an expanding spiral, demonstrating the outward development of a child into adulthood wearing a pair of ruby slippers. The cowardly lion believes he has no courage (politicians), the scarecrow that has no brain (population) and the tin man without a heart(repressed workers); this is the 3 components of the initiates within the mystery schools of ego, self, and soul.

Numbers

In the movie the *Matrix* 1999 Thomas A Anderson's passport was shown in a scene with the expiry date of expiry of 11th September 2001, the infamous date of 911. People who investigate the esoteric world of numbers know this. We will explore further; on the same passport it states the city of issue as CAPITAL CITY. Capital city in English Ordinal is 3+1+16+9+20+1+12+3+9+20+25 = 119, the inverse of 911. Here are some others:

- Death Ritual (EO) = 119
- All Seeing Eye (EO) = 119

In the movie, the elevator building explosion took place on level 41. Now let us look at the word 'fortyfirst' which is 6+15+18+20+25+6+9+18+19+20 = 156. Why is this a problem you may ask? The 156th prime number is in fact 911.

Do you think this is just a coincidence? In the movie *Die Hard with a Vengeance,* the hospital room clock has the time 10.10 and at the station says 9.11.

Secret societies use what they call 'Master Numbers'. When calculating numerological profiles, the digits of larger numbers are repeatedly added until they are reduced to a single digit up to 11. Instead of being added again which would leave us with 2, it remains 11. This is also the case with 22 and 33.

Number 47 is a favourite of JJ Abraham (the film director), he is a member of the 47 Society (yes that does exist). The number is used in *Fringe, Alias, Super 8, Star Trek, Star Wars: The Force Awakens,* and it is one of the Ark numbers in in the series *Lost.* 4+7=11

There are other movies where numbers replace letters, and these are:

- *SE7EN, Step Up 2: The Streets, The Thir13en Ghosts, Lucky Number S7even, 2 Fast 2 Furious, S1mOne, Pokémon 4Ever, L4YER CAK3, Leprechaun: Back 2 Tha Hood, Halloween H20: 20 Years Later, 5nal Destination*

Clones

The duality of man is mentioned elsewhere in his book with Janus worship, but we can explore cinematography over the years and what has clones associated with them:

1971 *The Resurrection of Zachary Wheeler*
1973 *The Clones*
1976 *Futureworld*
1978 *The Boys from Brazil*
1978 *Invasion of the Body Snatchers*
1978 *The Lucifer Complex*
1979 *Parts: Clonus Horror; The Dark Side of Terror*
1985 *Creator*
1993 *Jurassic Park; Body Snatchers*
1995 *The City of Lost Children*
1996 *Multiplicity*
1997 *The Lost World: Jurassic Park; Johnny 2.0; Alien Resurrection*
1998 *Mr. Murder*
1999 *Austin Powers: The Spy Who Shagged Me*
2000 *The 6th Day; The Other Me*
2001 *Replicant; Jurassic Park III*
2002 *The Adventures of Pluto Nash; Solaris; Star Trek: Nemesis; Star Wars: Attack of the Clones (Episode II); Repli-Kate*
2003 *Blueprint; Code 46; It's All About Love*
2004 *Appleseed; Godsend; Able Edwards*
2005 *The Island; AEon Flux; Invasion; Star Wars: Revenge of the Sith (Episode III)*
2006 *The Prestige*
2007 *Resident Evil: Extinction; Battlestar Galactica: Razor*
2008 *A Number; Star Wars: The Clone Wars; The Clone Returns Home*
2009 *Moon; S. Darko; Surrogates; Splice*
2010 *Resident Evil: Afterlife; Womb; Never Let Me Go*
2012 *Resident Evil: Retribution; Cloud Atlas; Universal Soldier: Day of Reckoning*
2013 *Oblivion; Orphan Black*

2015 *Splitting Adam*
2016 *Resident Evil: The Final Chapter; Morgan*
2017 *Logan; Kessler's Lab*
2018 *Replicas*
2019 *Us, Gemini Man; Living with Yourself; Pandora*
2020 *Horse Girl; LX 2048*
2021 *Seo Bok; Oxygen*

On the 5th July 1996, a domestic sheep (Finnish Dorset) called Dolly was birthed at the Roslin Institute in Midlothian, Scotland. The strange thing about this mammal was that it was the first mammal to be cloned from an adult cell and she went on to have 6 lambs (Bonnie, Sally, and Rosie (twins), Lucy, Darcy, and Cotton (triplets)). Dolly was euthanised at the age of 6 due to progressive lung disease that was not present in her original copy. The single cell that was used to create her came from a single cell of another sheep's mammary gland. The title of the front of *Time Magazine* which reported on Dolly was '*Will There Ever Be Another You (ewe)*'. In 2017 another pair of clones were created by the Institute of Neuroscience of the Chinese Academy of Sciences in Shanghai, but this time it was cynomolgus monkeys.

We know that secret government programs are always two- or three-decades (3×10) ahead of open market civilian, and public projects. Those of you who are research driven look up *Time Life Magazine* February 1965. Is this more predictive programming? If a lamb, why not a man?

Revelation of the Method
The purveyors of control believe that silence is consent. Everything requires people's consent and by not reacting we are consenting, to acquiesce is to accept what is presented to us. Like the gods of ancient Greece viewing mere mortals for entertainment from Mount Olympus, we look at the electronic shrine as a medium to transmit truth to our very eyes. Televisions are listed under patent as purely entertainment devises, meaning anything you watch does not have to be fact.

There is entertainment which appears to predict the future and *The Simpsons* is a perfect example of this. As you run through the seasons and episodes below you will see some predicted future events:

No.	Season	Episode	Title	Prediction
1	2	18	*Brush with Greatness*	Beatles Ringo Star answering fan mail from decades ago, Fans in Essex received a reply from Paul McCartney 50 years previously
2	2	9	*Itchy & Scratchy & Marge*	Marge debates freedom of expression, later Russian campaigners voted on the Renaissance statue being clothed or not
3	2	4	*Two Cars in Every Garage and Three Eyes on Every Fish*	3-eyed fish in a river by a power plant. Years later 3-eyed fish in reservoir fed by water from a nuclear power plant
4	3	24	*Brother, Can You Spare Two Dimes?*	Homer's half-brother's invention. Later the Cry Translator app
5	3	14	*Lisa the Greek*	predicted Washington Redskins would win Super Bowl XXVI
6	4	6	*Itchy & Scratchy : The Movie*	Billboard advertising which was later used to promote Kill Bill Vol 1
7	4	8	*New Kid on the Block*	Suing for false advertising. Later this legal discourse has made it into courts
8	4	20	*Whacking Day*	Holiday involving killing as many snakes as possible. Now we have the Python Challenge annual event
9	5	10	*$pringfield*	A spoof on Siegfried and Roy show, German magicians were attacked by a tiger. 10 years on Roy Horn is mauled by tiger on stage
10	5	15	*Deep Sapce Homer*	NASA elects an average person into space. Later UK holds a contest to turn an ordinary person into an astronaut
11	5	19	*Sweet Seymour Skinner's Baadasssss Song*	Lunch lady used horse parts for student lunch, horse DNA in beef burgers
12	6	19	*Lisa's Wedding*	Showing wristwatch comms in the future. Smartwatches rolled out years later
13	6	19	*Lisa's Wedding*	Lisa makes a phone video chat. Predating FaceTime feature
14	6	8	*Lisa on Ice*	Memo translate incorrectly. Common errors people blame on autocorrect
15	9	3	*Lisa's Sax*	Reading a book on Ebola Virus, American Ebola outbreak years later.
16	10	1	*Lard of the Dance*	Get rich scheme by Homer in siphoning grease. Delinquents using it in real life
17	10	2	*The Wizard of Evergreen Terrace*	Blackboard equation. Higgs boson wasn't confirmed until years later
18	10	22	*They Saved Lisa's Brain*	Homer speaking to Stephen Hawking about a doughnut-shaped universe. Now a genuine theory with the three-torus model
19	10	5	*When you Dish Upon A Star*	20th Century Fox division of Walt Disney Co. 20 years later Fox sale underway
20	11	17	*Bart to the Future*	Lisa as president with predecessor Donald Trump. 16 years prior to Donald Trump as president
21	11	5	*E-I-E-I-D'oh!*	Homer's farming "tomacco". Years later fruit and vegetables near nuclear power plant
22	12	15	*Hungry Hungry Homer*	More of an inspiration for a real-life event regarding a minor league team
25	20	4	*Treehouse of Horror XIX*	Voting machine recording incorrect selection. Later in Obama 2nd term same thing happened with a machine
26	22	1	*Elementary School Musical*	Bengt Holmstrom appears on a betting pool 6 years prior to winning Nobel Prize
27	21	12	*Boy Meets Curl*	US mixed doubles curling team come from behind to win Gold against Sweden. In the 2018 Olympic US curling team defeated Sweden
28	23	22	*Lisa Goes Gaga*	Lady Gaga doppelganger performs a song suspended on air. 5 years later Lady Gaga descended the roof of the stadium in the Super Bowl
29	25	16	*You Don't Have to Live Like a Referee*	Arrested. A year later a FIFA official arrested on corruption charges
30			*The Simpson Movie*	NSA Spying scandal - Edward Snowden whistle

God's Telephone Number

Have you ever noticed that as you watch movies that there is a repetition with some telephone numbers?

In 1973 American area telephone codes changed, this meant that each state had a prefix of 3 numbers whereas before they were letters and numbers. This created problems when numbers were shown in movies as people would contact those numbers. The movie industry agreed with the telephone industry to use the prefix '555', meaning a fictitious area code. The next time you watch a movie, and an American telephone number is used you will notice that it starts with '555'. They use a range of numbers from 555–0100 to 555–9999. Here are some:

- 555-0134 was used in *Fight Club, Momento, Harriet the Spy, Someone Like You*, and the TV show *Millennium*
- 555–2368 was used in the *Ghostbusters*, appeared in *Momento, The Rockford files, The Mod Squad* and others

In the movie *Bruce Almighty*, God's phone number showed up on Bruce's pager as 776-2323. The producers had checked that the number did not exist where the film set was (Buffalo, New York) but the problem was they did not check other states across the country which meant that wrong numbers and prank calls were made requesting to speak to God. A minister for a church in North Carolina also received calls wanting to speak to God, it was only a coincidence that the minister's name was Bruce. This number was purchased by the studio with an apology.

As a child I grew up watching the *A-Team* and Hannibal's number was 555-6162 and people still to this day ring this number seeking assistance from the commando team, the soldiers of fortune.

Recently production companies are using real numbers, and which allows for a message from the actors associated with the show. *The Office*, *Breaking Bad* and *Stranger Things* are example of these. *The Matrix* the opening cipher traces the telephone number used by Trinity as 312–555–0690, if dialled you get a message from *The Matrix*. In the TV series *Breaking Bad* (S1, E2) Jesse Pinkman a freestyle rapper and Crystal Meth chemist leaves the following message on his answerphone *"Yo, yo, yo! 1-4-8-3 to the 3 to the 6 to the 9, representin' the ABQ. What up, Biatch? Leave at the tone"*, if you ring his number you still get this reply.

Even in the UK this goes on, as production companies always have the area number 01632. Next time you watch anything with a telephone number, now you know the occult used.

911 to phone the police

A red room number is one to dial with caution. It is also referred to as a *red rum* number and if you are a follower of the movie *The Shining* you will understand *red-rum* spelt backwards. Allegedly, it can be used to locate the physical location of people wanting to commit suicide who are offering themselves to be hunted, kidnapped and even killed by being taken to a "red room". These alleged torture sessions and murders are said to be broadcasted live over the dark web. These numbers have demonic voices and terrifically uncomfortable and leaving your number is not worth venturing. One such number is a US area code 408-634-2806, if you fancy calling. Since I found these numbers whenever anyone ask me if they can use my phone, I tend to say it is broken (if it is the landline) or out of credit (for my mobile). Though if I am in an office with people, you do not like, the possibilities are endless.

Whenever I discuss dark stuff, I try to leave the reader on a brighter note. Wrinkles is a clown from Naples in Florida. Nothing odd in that, but this clown will *'prank your friends'* according to his web site, and even scare your kid *'straight'* whatever that means. When I contacted him, he was a creepy guy in his 60's and says he is from Rhode Island in the States. He seemed very professional in the *'scare the crap out of you'* clown market. Call and leave a message and he will phone you back, he did me. His quote was expensive to stalk my best friend for 3 days, it included a hotel in the day (he works nights), the flight to the UK and expenses (he uses blood in his work). It worked out at £7,300 but he does take bitcoin, so he is tech savvy. Call him on (407) 734-0254 but please get the number right, I misdialled the first time so you can only imagine the message I left on some poor American housewife's voicemail, she never phoned me back, nor me to her. Stuart you are a lucky I am not a wealthy man.

Chapter 17 - Giants and Champagne

Some of our champagne bottle names are based on the Bible and the size of giants, let us court this idea.

Name	**Size X Bottle(s)**	**Glass Size**	**Meaning**
Piccolo	¼ of a bottle	1	Italian (for small) or child
Half/Demi/Fillette	1/2	3	French (meaning half or little girl)
Imperial Pint	600ml	4	Rare by Pol Roger, associated with Sir Winston Churchill; he preferred the Pint-size bottle as it ideal
Bottle/Standard	1	6	Standard bottle, average size
Magnum	2	12	Latin (*great*), larger size
Jeroboam	4	24	1st King of the Northern Kingdom of Israel (Meaning *he increases the people*)
Rehoboam	6	36	Son of King Solomon, Biblical King of Judah (meaning *he who enlarges people*)
Methuselah	8	48	The aged one (in the Bible he lived for 969 years) Genesis 5:27
Salmanazar	12	72	One of the 5 kings (of Syria), Assyrian King Salmanazar III
Balthazar	16	96	King of Arabia (1 of the 3 wise men at the birth of Jesus)
Nebuchadnezzar	20	120	Babylonian (powerful King) Daniel 3:1-7, King of Chaldeans made an image of gold, whose height was threescore cubits, and the breadth thereof six cubits: he set it up in the plain of Dura, province of Babylon
Solomon	24	144	King of Israel (or son of King David)
Sovereign	34	200	Taittinger created it to celebrate the opening of the largest (at the time) cruise ship, *Sovereign of the Seas* in 1988
Primat/Goliath	36	216	Latin (*primas* – chief, noble) Goliath was the giant Philistine warrior that David defeated in the biblical story
Melchizedek/Midas	40	240	King of Salem Genesis 14, he died at the age of 465 years old. He was a king and a priest who lived a very righteous life in the Old Testament. It is sometimes called Midas after the Greek mythical king who turned everything he touched to gold Today only produced by Champagne Drappier

While this is a fun note, we celebrate using this fizzy liquid, the occult involves old age and the possibility of colossus humans. Let us go deeper.

Ancestral Mathematics and Giants

The body of work I have created here is a hypothesis based on proof supported by theory with syncretic research. This strives to unite people, reveal secrets and most importantly open discussion about our true reality.

For you to have been born you need to stand on the existence of ancestors. Let us collate the numbers into a pattern.

2 parents, 4 grandparents, 8 great grandparents, 16 second great grandparents, 32 third great grandparents, 64 fourth great grandparents, 128 fifth great grandparents, 256 sixth great grandparents, 512 seventh great grandparents, 1024 eighth great grandparents 2,048 ninth great grandparents, and here is the pattern:

great, great, great, great, great, great, great, great, grandparents
great, great, great, great, great, great, great, grandparents
great, great, great, great, great, great, grandparents
great, great, great, great, great, grandparents
great, great, great, great, grandparents
great, great, great, grandparents
great, great, grandparents
great, grandparents
grandparents
parents
YOU

The problem here is that this pyramid theory of doubling the number of ancestors in each generation creates an inverted pyramid, eventually the world's population would not be large enough to support the numbers and therefore we would all be related in family.

There is something that happens every so often, let us just call it *'a great reset'*. If we have an expansion of societal numbers, we must taper back the number of grandparents according to how many souls the planet can support or in history what estimated to be alive.

I have calculated for example of 12 generations = 2048, generation 13 = 4096, generation 14 = 8192, generation 15 = 16384, generation 16 = 32768 and it creates a diamond shape.

You
2 parents
4 grandparents
8 great grandparents
16 great, great grandparents
32 great, great, great grandparents
64 great, great, great, great grandparents
128 great, great, great, great, great grandparents
256 great, great, great, great, great, great grandparents
512 great, great, great, great, great, great, great grandparents
1024 great, great, great, great, great, great, great, great grandparents
X-G–G- great, great, great, great, great, great, great grandparents
X-G–G–G–G–G– great, great, great, great, great grandparents
X-G–G–G–G–G–G–G–G– great, great, great grandparents
X-G–G–G–G–G–G–G–G–G–G-G-G great grandparents

Why does it turn inwards, quite simply it is cousins who married cousins until the population grew to its maximum, in the above example see line 1024. Wars, plagues, natural disasters thinned out the population. I believe there is an inverse thinning going on now.

Historically some cultures encourage cousins and close relations to marry for reasons of bond, wealth retention, and property. Many royal and banking families have carried out this practice (*endogamy*). However, there is a genealogy issue here and it is called *'Pedigree Collapse'*.

There is a calculation called the *Square Cube Law*. The theory is that if the genes of individuals in a family line are consistently repeated through their offspring, the decedents will get smaller and smaller. If the above diamond theory works, it means giants did indeed walk the earth. The square-cube law is the deceptively simple observation that as you scale up dimensions — say, as you get taller — then area of the object increases proportional to the square of the length, while volume increases proportional to the cube of length. While no scientific proof exists, it would seem any discussion on the subject is shut down. The Smithsonian Institute hold all scientific skeletal findings on this matter which is not open to the public.

Time

To be born today from 12 previous generations, you needed a total of 4094 ancestors over the last 400 years. Not to mention the battles, difficulties, sadness, happiness, love stories, and hope that happened to you to be here reading this book. You are truly remarkable. Your understanding of how you can manifest a positive position for yourself and indeed a negative one, is a state of mind and a frequency of resonance.

Mathematically a generation is from when a female is born to when she first gives birth, which 100 years ago was 15. This number, however, has changed since females choose to wait until later in life to give birth. A generation as of 2022 is 26 years.

Let me leave you with a few questions...were we giants? did we live much longer? and have resets happened periodically through history?

Chapter 18 - Blood, Zombies and Plimsoles

During my lifetime I have occupied many properties both rented and mortgaged (*Death Pledge)* and have always redecorate upon taking occupancy to have a comfortable existence. I do, however, have a wicked sense of humour and always hand over the keys to the new occupier/owner with a wry smile on my face. While redecorating and prior to re-papering the walls I use a tin of blood red coloured paint and on the largest wall write the following '*I will kill again*'. I allow the paint to drip down the wall and dry, I re-paper and decorate. I know that at some point down the line someone will strip the paper and find the message, a pity I will not be there to see their faces when they discover the words and the red arrow pointing towards the garden!

Quran of Blood

On 8th April 1998 Saddam Hussain commissioned a Quran which bizarrely was said to be entirely written in his blood. The pages soaked up a total of 27 litres of haemoglobin for the 114 chapters and 6,000 verses, the 605 pages was completed in the year 2000. On the numbers this may seem innocuous but in the West an individual can legally give only 6 units or 3 litres of blood in a year (12 month) so by this reckoning it would have taken 9 years to sustain a healthy immune system. The Muslim faith consider anything touched by blood as unclean and *haram* or forbidden. Touching a canine is haram which is why Muslim taxi drivers in the UK have been prosecuted for refusing fares from people who have guide dogs. Inversely cats are not and are allowed in mosques. Qurans like Bibles should not be defaced (written in) as it is considered blasphemy.

When Iraq fell to the coalition forces in America and Britain's phony war, the book created a dilemma for clerics of Islam. Presently it is not on show in a museum, but clerics are still considering what to do with it and whether to hold it in places that demonstrate totalitarianism such as the Hitler or Stalin Museums.

Another Lost Book

Like the Book of Enoch which I visited in *stopper 1,* the Book of Jasher also being called Pseudo-Jasher is being heavily scripted as a hoax by mainstream researchers and search engines. The origin of this translation, however, seems to stem from November 1750 which claims to have been written by Jasher, son of Caleb a lieutenant of Moses and a book which is mentioned in the Bible.

These parchments have some remarkably interesting verses and I draw your attention to (verse 29) *"and his wise men and SORCERERS said unto him, that if the blood of little children were put into the wounds he would be healed"*. (verse 30), *"And Pharaoh hearkened to them, and sent his ministers to Goshen (*where the Hebrews lived*) to the children of Israel to take their little children"*. (verse 31) *"And Pharaoh's ministers went and took the infants of the children of Israel from the bosoms of their mothers by force, and they brought them to Pharaoh daily, a child each day, and the physicians killed them and applied them to the plague; thus did they all the days"*. (verse 32) *"And the number of the children which Pharaoh slew was three hundred and seventy-five"*. The pharaoh despite his insanity died of skin boils and infections. At this point remember the word SORCERERS mentioned in verse 29, it is the same word in the chapter on Coronavirus and the Bible.

The period of child stealing, killing and blood drinking was under Thutmose III (6th Pharaoh of the 18th Egyptian Dynasty). This occult policy of child abuse and blood consumption resulted in the great exodus. Interestingly when we turn to Revelations 16:1 *'Then I heard a loud voice from the temple saying to the seven angels, "Go and pour out the seven (7) bowls of God's wrath on the earth"'*. 16:2 *"The first angel went and poured out his bowl on the land, and ugly, festering sores broke out on the people who had the mark of the beast and worshipped its image".*

Interestingly the Book of Jasher was endorsed by the Rosicrucian Secret Society of San Jose, California in 1934 who also has the "secret of resonance" which is mentioned elsewhere in this book.

The Bible and Blood

Blood as a word is the 2nd most used word in the Bible after God. It is used 447 times in 375 verses. Now look at the numbers 4+4+7 = 15, 1+5 = 6 and 3+7+5 = 15, 1+5 = 6.

Look at specifics of where blood is mentioned in the book itself. Revelation 16:10 *"And the fifth angel poured out his vial upon the seat of the beast; and his kingdom was full of darkness; and they gnawed their tongues for pain"* 16:11 *"and blasphemed the God of heaven because of their pains and their sores, and repented not of their deeds".* 16:12 it continues *"Then the sixth angel poured out his bowl on the great river Euphrates, and its water was dried up, so that the way of the kings from the east might be prepared".* 16:13 *"And I saw three unclean spirits like frogs coming out of the mouth of the dragon, out of the mouth of the beast, and out of the mouth of the false prophet".* 16:14 *"For they are spirits of demons, performing signs, which go out to the kings of the earth and of the whole world, to gather them to the battle of that great day of God Almighty".*

In the 15th century Pope Innocent VIII's physician is said to have bled out 3 boys to death and fed the warmblood to his patient but alas 4 corpses were the result at the end of that fateful day.

The drinking of human blood by homosapiens is increasing year on year. The so called *sanguinarians* or blood drinkers (*med-sang*) believe this elixir extends life, improves life force, remedies fatigue, headaches, and stomach pain. These self-identifying vampires insist 1 pint per week sustains their healthy existence. Most major cities on the planet have these underground communities and their numbers are increasing. Once they have a willing donor, they go to great lengths to make sure the donors are healthy. Upon an agreed bleed if the recipient decides not to drink instantly, they add Sodium Heparin (anti-coagulant) as a preservative allowing to drink it later. A pint of blood has twice the calorific value as a pint of beer. Anyone who has a friend with a history purchase of phlebotomy kits may need a new circle of acquaintances. These are openly for sale on eBay accounts (they contain vacutainers, lancets, needles, syringes, tourniquets, and gauze).

Historically in Europe those afflicted by epilepsy could gather around the gallows and/or guillotines to collect the warm blood to drink or make soups, it was called *corpse medicine*. The elixir was thought to absorb the dead's energy thereby curing those who were afflicted.

Here is a word to add to your vocabulary, *"hellbroth"* the definition is 'a broth prepared with magical intent (in black magik) for an evil purpose of invocation of thelemite anti-Christian doctrine' and was first to be used in documents from a 1616 dictionary. On page 127 *The secret society rituals of the Ordo Templi Orientis* (O.T.O) has bloodletting as one of its degrees. The origin of blood drinking is from the Chassidim, sacrifice of children and more recently in the reprinted works of Dion Fortune the for-runner of Allister Crowley. William Shakespeare even got in on the act (Scottish play or *Macbeth*) Act 4 Scene 1 *"For a charm of powerful trouble, like a hell-broth boil and bubble".*

Shoes

In *stopper 1* I mentioned the red shoe clubs but let us turn to numbers on footwear.

At 55 I am of older age and as child I wore plimsoles for exercise, for those that are younger these are now called sneakers which at times are altered for rarity. For those who are older, a custom shoe is new footwear hijacked by other companies and stylised independently. A bit like taking my old Land Rover and giving it to my local garage for speed stripes, *'cos that is how I roll'.*

In 2019 creative label MSCHF customed a shoe called the "Jesus Shoes", they were advertised as the Air Max 97s and said to endorse the miracle of Jesus walking on water. They had a crucifix on the laces and the cushioning in the back sole is detailed and viewable tube filled with holy water from the River Jordan. Biblically if you look at the number stitched onto the lateral forefoot is Matthew 14:25 KJV Bible *"and in the fourth watch of the night Jesus went unto them, walking on the sea".* Jesus' walking on water was not lost on me and their bravery not to reference other religious groups left them unscathed.

Why was the sneaker white and baby blue? Well, let us not forget (Freemason 2nd and 3rd degree use these colours) and they sold quite well over a week period. At 3000 dollars a pair they sold out in 7 days.

Interestingly in Reverse Full Reduction *'Jesus shoes'* adds up to 58 but so does the unluckiest number of misfortunes in the mystery schools.

Look at some ordinal numbers:

- *Secret society* (FR) 1+5+3+9+5+2+1+6+3+9+5+2+7 = 58
- *Rosicrucian* (FR) 9+6+1+9+3+9+3+3+9+1+5 = 58
- *Freemasonry* (FR) 6+9+5+5+4+1+5+1+6+9+7 = 58
- *Gregorian* (FR) 7+9+5+7+6+9+9+1+5 = 58
- *Solomon's temple* (FR) 1+6+3+6+4+6+5+1+2+5+4+7+3+5 = 58
- *Annuit Coeptis* = 58 (the Latin statement on the US dollar bill)

The Rosicrucian Order means *'rose cross'* but if you take these words under Reverse Full Reduction 9+3+8+4+6+9+3+8+8 = 58. These coincidences continue, take the Icon Crucifixes which has a scroll on the base under the feet of Jesus with 4 letters, *INRI*. This stands for *"Iesvs Nazarenvs Rex Ivdaeorvm"*. Latin uses *"I"* instead of *"J"*, and *"V"* instead of *"U"* (so you would get *Jesus Nazarenus Rex Judaeorum*). In English it would be "Jesus of Nazareth, the King of the Jews."

Take INRI in (RO) 18+13+9+18 = 58 and now you can see the hatred of this number by the mystery schools. The King James Bible was named after himself and to make you aware he died unexpectedly and guess at what age? 58.

The well informed amongst you will be aware of Lil Nas X a rapper from America. I listen to all music genres' and I must confess I liked one of his songs, not anymore. Now take LIL n rapper (FR) 3+9+3+5+9+1+7+7+ 5+9 = 58.

The rapper and influencer endorsed a custom shoe which was a collaboration with MSCHF and was advertised as *'Satan Shoes'*, total opposite to the *'Jesus Shoes'*. They were jet black and referenced a biblical text but this time it was Luke 10:18 *'And He said to them, "I saw Satan fall like Lightening from heaven"'*. The number 10:18 continues as they RRP was $1,018 a pair. If you take *"a Satan's sneaker 666"* (FR) 1+1+1+2+1+5+1+1+5+5+1+2+5+9+6+6+6 = 58. If you were to RFR the words *'black shoes'* you also get 7+6+8+6+7+8+1+3+4+8 = 58. They do love these numbers.

As a side note the same month these plimsoles came out in the US, the estate of Dr Seuss decided to cease publishing 6 books by the children's author and a week later the toy Mr Potato Head was rebranded. 'Mr' was removed from the name both products being considered as too politically incorrect while "WAP" (wet-arse pussy) won the night at the Grammys; humans have truly fallen from grace but back to blood.

If you want Satan under your feet, then too late as the 666 available pairs have sold out (they sold out online in one 1 minute). I would not want a biohazard under foot, if they sprang a leak with blood all over the place it would be considered a Hazchem incident, right? Well, you would wrong. We are surrounded by blood, how so? Just keep reading.

Soylent Green and Showers

If you are old enough like me to remember the dystopian movie from 1973 with Charlton Heston, then you will know that at as he is dragged off strapped into a gurney he screams *"Soylent Green is people"*. The number 300 is important because alkaline hydrolysis needs this temperature in degrees (F) to dissolve the human body into a soup, sometimes called a 'green cremation'. I hope your sitting comfortably for the next facts. In the West immediate burial of loved ones is not required and open caskets allowing viewing of the deceased is normal. To achieve this the body must be preserved otherwise decomposition starts within 8 hours after death. Have you ever thought about the embalming process when they pump the cadaver full of preservative? Well, you should. For every 50 to 75 lb of body weight it takes about a gallon of embalming solution made up of formaldehyde emptying the body of blood. This biological fluid ends up in your kettle. NO! I hear you scream. The preparation of an expired individual who is filled with formaldehyde has the resultant blood forced from the body. Where does this blood go? Well, it is flushed down the drain into our sewers. Medical waste, swabs, bandages, cotton dressings and clothing which contain blood are treated as a biohazard by hospitals, but blood from the body in an undertakers or mortuary is a non-issue. Once flushed down the sinks and drains it re-enters the water cycle system once the water treatment plant makes it all nice and sends back through your taps for cooking, drinking, bathing all in the essence of our ancestors, relatives or even the neighbour you did not like.

In *stopper 1* I explained that water has energy and on the calculations 0.0000000000033% of drainage water contains blood. The number 33 appears and if you take "blood in waters" (FR) 2+3+6+6+4+9+5+5+1+2+5+9+1 = 58.

Pepsi - other products are available

In 2019 I discussed on a talk with the Alternative View Network human consumption and the meddling by scientists. In 2010 PepsiCo signed a deal with a biotech firm Senomyx for a new sweet taste technology. To be clear Pepsi-Cola (PepsiCo) have said they do not conduct or fund any research that uses tissue or cell lines (mediums to grow things in) derived from human embryos or foetal cells. Now let us explore the language, in the 1970s cells were taken from the kidney of an aborted baby and given the prefix HEK-293. This cell was one application to help mimic taste receptor cells in the development of possible new flavours for humans. HEK-293 which is used today is not directly from the cell line but is a clone of the original cell. The company dropped their involvement with Senomyx. While you do your research on this consider that if these scientists had continued, we would have '*Pepsi HEK* 293' on the market. In (FR) this is 7+5+7+1+9+8+5+2+2+9+3 = 58. I am old enough to remember the Pepsi challenge which was put back-to-back against Coca-Cola and I in fact preferred Pepsi.

Beauty Cream

The need for people who are older to look younger is a billion-dollar industry. In 2015 the HuffPost revealed a new "baby foreskin" cream by SkinMedica. The product uses foreskin fibroblasts grown from penis stem cells from babies. The original research was to explore skin growth for burn victims, diabetic ulcers, and other life-saving applications. Skin is the largest organ in the human body but stretchable skin i.e., from the penis of male babies is a cell line due to blood supply levels that can only found in this area. This vampire facial cream (as I call it) made the audience at one of my lectures extremely uncomfortable and so it should. Protecting babies and children is the one essential thing a human can do and to mutilate the genitals of girls is illegal but ask yourself why boys are up for grabs in the desperate hope to look younger. I am growing old gracefully accepting there is always a younger model, why cannot other people do the same and leave babies alone.

Nimrod and DNA

During my research over the last two decades, I have found a plethora of information regarding the history of humanity going back to Sumeria and ancient Mesopotamia, which in turn has lines and tales in mythology and legend. In 2003 the Iraqi government declared that a German led expedition had discovered the lost tomb of King Gilgamesh in Uruk, deep in the Iraqi dessert. An announcement page by the BBC dated Tuesday 29th April 2003 at 7:57 GMT reaffirmed it. This grabbed my attention for the simple reason that Gilgamesh was mythical and there are theories he might have been Nimrod. 15 years ago, I came across a document which was translated into English having been written 2100 years before the birth of Christ and it was called the *Epic of Gilgamesh*. *Gilgamesh we have his DNA* (FR) 7+9+3+7+1+4+5+1+8+5+5+8+1+4+5+8+9+1+4+5+1 = 101.

The depiction (713- 706 BC) held in the Louvre of Gilgamesh holding one of his pet lions.

I looked at the Magnetogram mapping (as a surveyor I use this), its location and city size raised another fact. Historically it seems to have been a kind of Venice in the desert but its location according to historical maps indicate it was on the former Euphrates River. This in turn implies that it was in fact on the site of where the Tower of Babel was constructed. Biblically this tower (for those who are aware of the passage within the Bible) was an affront to God, humanity was seeking deferment to seek the creator by building vertically to heavens. The creator according to this biblical text did not take too kindly to mankind's actions and the Gilgamesh destroying it, casting humanity into locations all over the world and giving them different languages to confuse them so that this would never be repeated.

Numbers continues for Gilgamesh as he was (apparently) 1/3 god, 1/3 giant (*Nephilim*) and one 1/3 man. 1/3 is of course one part of 3, so his genetic make-up according to these myths was 333% x 3 = 100% but no. If you take 100 and ÷ by 3 you get 33.3333333 to the infinite. But in reverse if you take 33.33333×3 = 99.99999 to the infinite. The 3, 6 and 9 are here again. This DNA pattern shows some fascinating numbers for If indeed he was 2/3 genetic and 1/3 god it would indicate that his earthly presence at 2/3 would be 66.6666666 to the infinite. 2 ÷ 3 = 0.666666. If you divide this by 212° Fahrenheit or the boiling point of water and bang 31446 appears, do you not see it. Pi or 3.14, the number of the building blocks of life appear.

There is a claim that a document dated 12/13/2018 ref F2019–02110 to U.S. State Department by Denetra D Senigar states "requesting documents pertaining to the resurrection chamber of Gilgamesh also known as Nimrod, the location of his body and the location of the buried Nephilim (giants)". This freedom of information request was to Hillary Rodham Clinton and has since disappeared but there are ways of finding it. Consider these 2 facts, the Iraqi invasion took place between 20th and 1st May 2003, and the Gilgamesh Discovery was on Tuesday 29th of April 2003 in the mainstream media. The American charge headed straight for the Iraqi National Museum, in the guise of protecting historical artefacts. To this day all references to Gilgamesh have disappeared. If you have a body, you have DNA and subsequently gene evidence of a Nephilim and we know how insane scientists like to mess around with this stuff.

Rh negative blood and dolphins

I need to point out that I have type O blood. Years ago, I had a motorcycle accident and had to have an Xray of my spine, to my shock the radiologist commented that I was type O just by looking at the x-rays. I asked him how he knew, and he said that I had 34 vertebrae not the regular 33 typical of other blood types. Interestingly I also have all the other aspects of O; two different colour eyes, my body repairs extremely quickly and my IQ is above average. The other term used is *Dragons Bloodline*, all Dolphins and Turtles have O Rh-negative blood, and the ninja turtles were mutants if you remember. Within primates there are 612 species and sub species recognised by the International Union for Conservation of Nature (IUCN) and not one has Rh-negative, they are all positive. This in turn suggests in Darwinism Theory that mankind is built on this chassis so what is the hypothesis of these Rh-negative humans. One thing I can declare I would rather be a dolphin then a dragon.

The Rh system blood groups in humans are:

- A RhD positive – (A+) one of the most common blood types
- A RhD negative – (A-) rare blood type, less than 10% have this blood type
- B RhD positive – (B+) rare blood type, less than 10% have this blood type
- B RhD negative – (B-) rare blood type, less than 10% have this blood type
- AB RhD positive – (AB+) "universal recipient"
- AB RhD negative – (AB-) rare blood type
- O- is the rarest of them all

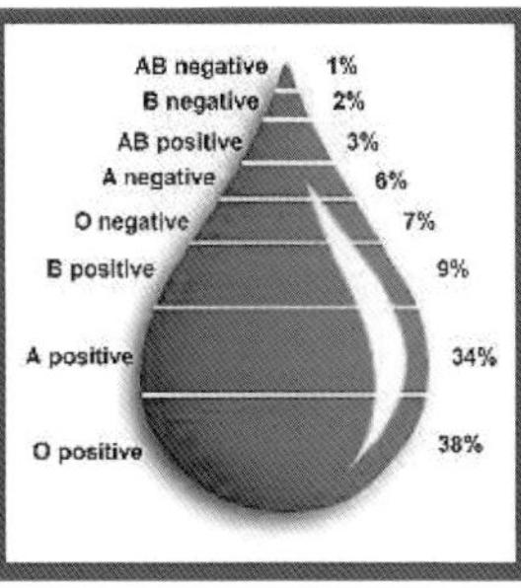

If Your Blood Type Is...

Type	You Can Give Blood To	You Can Receive From
A+	A+ AB+	A+ A- O+ O-
O+	O+ A+ B+ AB+	O+ O-
B+	B+ AB+	B+ B- O+ O-
AB+	AB+	Everyone
A-	A+ A- AB+ AB-	A- O-
O-	Everyone	O-
B-	B+ B- AB+ AB-	B- O-
AB-	AB+ AB-	AB- A- B- O-

"If mankind evolved from the same African ancestor then everyone's blood would be compatible, but it is not. Where did the Rh- negatives come from? Why does the body of an Rh- negative mother carrying an Rh+ positive child try to reject her own offspring? Humanity isn't one race, but a hybrid species."

Species with Amnesia: Our Forgotten History by ROBERT SEPEHR

85% of the world population has Rh-positive blood. Interestingly, Royal Families have O negative blood. Rh-negative humans have a higher IQ, sensitive vision, lower body temperature, high blood pressure, increased occurrence of intuitive abilities, prominently blue green or hazel eyes, reddish hair, and an increased sensitivity to heat and sunlight. Additionally, a Rh-negative mother carrying a Rh-positive child mostly reject the pregnancy and vice versa, the positive and negative mixed results in the rejection of a child.

Redheads and the spike protein

In 2004 a study was published in *Anaesthesia Magazine* revealing that red-haired people require up to 20% more anaesthetic to achieve the same results both in general and local anaesthetic during a medical procedure. Without boring you with the science, red-headed people and people with Rh-negative blood have mutations on the MC1R gene that not only increases the expression of red pigment but may also alter the function of the central nervous system. I suffered this reaction during an operation on my foot, I woke up halfway through the procedure and this resulted in the surgeon having to use local anaesthetic to complete the surgery. Studies have been carried out that shows Rh-positive people are far more susceptible to spike protein issues than those who have Rh-negative blood (MedRxiv2020, DOI document 1.1101 - O rhesus negative blood). Behold document number 1.1101 or 1111. It seems the O rhesus negative people on the planet are those that are less infected, all have issues with an introduced spike protein. It therefore looks like the Royal family will be fine having a form of natural immunity.

Nostradamus

Is hailed by many to have predicted the death of kings, the Great Fire of London in 1666, the start of World War II but there is another thing I came across within his readings. *"Few young people: half-dead to give a start. Dead through spite, he will cause the others to shine, and in an exalted place some great evils to occur, sad concepts will come to harm each one, temporal dignified, the Mass to succeed. Fathers and mothers dead of infinite sorrows, women in the mourning, the pestilent she-monster: the Great One to be no more, all the world to end".*

Ships and Blood

In the *stopper 1*, I mention banking, shipping, and water terms, however, there are more deep-seated elements with shipping. Have you ever wondered why ships are christened as humans are in the West? This is a ceremony of rite of passage, (*shipwright* – builder of vessels for passage). Biblically in Psalm 107:23 lines I mentioned to protect Mariners.

"They that go down to the sea in ships, that do business in great waters; These see the works of the LORD, and his wonders in the deep".

To christen a ship or seagoing vessel (Ephemera Ceremony) is to give it luck for safe passage while sailing. Christening of a vessel by any military or government organisation can only be carried out by a female. But what has this to do with blood I hear you ask. While launching a new vessel the Babylonians sacrificed an ox, the Turks sacrificed sheep, the Ottomans used sheep to Allah, while the Tahitians, Viking and Norse offered up something quite different. The latter would obtain an enemy, dispatch him, use his skull to drink from thus the origin of the word Scholl (*Skol/Skål* - cheers or good health), when making a toast after flaying the body across the bow of the new vessel and pouring blood down its figurehead. This toast of christening continues today with the smashing of a bottle of champagne across the new vessel while naming the ship, it is a bloodletting ceremony. Fancy a cruise anyone?

Skateboards

'*Liquid Death'* skateboards mixed composite paint with the blood from Tony Hawk the world's most famous skateboarder. If you look at the release dates of the 'Satan Shoes' (Monday 29th March 2021) and the 'Liquid Death' (Tuesday 24th August 2021), it is a total of 149 days which is the 35th prime number. The number 35 is important as you can see from (FR) Catholic = 35, Jesuitism = 35, Baphomet = 35, in (RFR) Satan = 35 and in (FR) Blood = 21, Evil = 21, Jesuit = 21. Agenda 21 anyone?

A body farm is an extraordinary place, in hidden locations they exist all over the world and share findings about natural decomposition of a human. Set in various climates, the bodies (donated hopefully) are arranged in scenarios such as in the boot of a car, in water, under fences, on stone slabs, in a pit after a fire or in a concealment setting. In 2019 researchers at the 1st body farm to be set up in Australia (Researcher Alyson Wilson) found that the dead bodies move during decomposition. Professional investigators are aware that bodies do not necessarily remain in the same position when life leaves a corpse, thereby altering some aspects of a homicide investigation. The movement it seems is down to dehydration of ligaments, thereby moving limbs and on occasion with extraordinary results like bodies rolling over which gives a whole new dimension to turning in your grave. On the rare occasion exhumation has taken place, it has historically shown that some bodies have been found facing down in coffins. This gave rise to many urban myths for the dead. The Australian Facility for Taphonomic Experimental Research was not lost on me as its facilities acronym is AFTER.

Zombies

CONPLAN 8888-11 is a Pentagon document *"Counter-Zombie Dominance"*, it detailed a battleplan against zombies for the US military command. Since its discovery, the military have released counter documentation stating that it was for training of recruits at strategic command.

All cultures have legends/folklore/mythology for zombies:

- Revenant (European), Draugar (Norse), Ghoul (Arabic), Jiangshi (Chinese), others are Walkers, Undead, Shufflers, Risen, Munchers, Zeds, Turnskin, Crawlers, Saifu, Skels, Z's, Undead.

I have studied in detail the different holy books, the Bible, the Quran, the Torah and many other ancient teachings manuscripts and texts as far back as the Sumerian Tablets. The one single thing that crosses my mind is that these books are a type of warning to humanity regarding their behaviour. I came across a remarkably article last year which was about the blog post by the US Centres for Disease Control and Prevention (CDC) and it was called *"Preparedness 101: Zombie Apocalypse"* published in 2011. Governments must always have contingency plans for worst case scenarios for assorted reasons such as civil unrest, force majeure (Latin - *act of God*), earthquakes, tsunamis, the list is endless. But in this instance, it alarmed me as the document expressly discusses preparedness for a zombie apocalypse. I hear you say...this cannot be a real thing...preparedness against the undead...look it up yourself. The *Walking Dead* TV shows writer revealed the cause of the zombie apocalypse started in Atlanta, Georgia which so happens to be the home of the CDC which was established on 1st July 1946. *I am legend* another zombie movie coincidentally was released on 14th December 2007, the reason this is interesting is that it is exactly 13 years to the day that vaccines for coronavirus went live in the United States on 14th December 2020.

Over the year's zombie movies have increased tantamount, drowning of entertainment with the initiation of such biblical outbreaks. This can be anything from radiation mutations, VIRUSES, mad cow disease, measles, and rabies.

Zombie movies year on year have been ramping up and here are some:

1941 *King of the Zombies*
1943 *Revenge of the Zombies*
1964 *Children of the Living Dead*
1966 *The Plague of the Zombies; The Frozen Dead*
1968 *Night of the Living Dead*
1971 *The Omega Man*
1972 *Dead of Night*
1973 *The Crazies*
1977 *Rabid*
1978 *Dawn of the Dead*
1979 *Zombie Flesh Eaters; Zombi 2*
1980 *Hell of the Living Dead; City of the Living Dead; Zombie Holocaust*
1981 *Night of the Zombies*
1982 *The Treasure of the Living Dead*
1985 *Re-Animator; Hardrock Zombies; Return of the Living Dead*
1986 *House*
1987 *Evil Dead 2*
1988 *Zombie 3*
1989 *Pet Cemetery; Redneck Zombies; The Dead Next Door*
1990 *Night of the Living Dead*
1992 *Army of Darkness; Braindead, aka Dead Alive*
1993 *Return of the Living Dead III*
2000 *Versus*
2002 *28 Days Later; Resident Evil*
2003 *Undead; House of the Dead*
2004 *Shaun of the Dead; Dawn of the Dead; Resident Evil: Apocalypse; They Came Back; Zombie Nation*
2005 *Land of the Dead; Severed; Tokyo Zombie; Evil; The Return of the Living Dead: Necropolis*
2006 *Fido; Slither; Black Sheep; Zombie Diaries; Wicked Little Things; Special Dead; Last Rites; Horrors of War; Mulberry St*
2007 *28 weeks Later; I Am Legend; Planet Terror; Resident Evil: Extinction; Diary of the Dead; Brain Dead; Wasting Away; The Mad; Dead Heist; Zombie Town*
2008 *Quarantine; Pontypool; Dead Girl; Colin; Dance of the Dead; Mutant Vampire Zombies from the 'Hood! ; Zombie Strippers; Reel Zombies; Night of the Flesh Eaters; Rica: The*

Zombie Killer; Ninjas vs. Zombies; Dead Eyes Open; The Dead Outside; The Vanguard; Day of the Dead

2009 *Zombieland; Dead Snow; The Horde; Autumn; Blood Moon Rising*

2010 *The Walking Dead; The Dead; Zombie Beach; 28 Hours Later: The Zombie Movie; The Book of Zombie*

2011 *Exit Humanity; Juan de Los Muertos; Deadheads; Remains; Eaters: Rise of the Dead; Humans vs Zombies*

2012 *[REC]3: Genesis; A Little Bit Zombie; Detention of the Dead; Dead Season*

2013 *World War Z; The Returned; Miss ZOMBIE; Warm Bodies; Contracted*

2014 *Life after Beth; Zombie Fight Club; Goal of the Dead; Zombeavers; Burying The Ex*

2015 *Scouts guide to the Zombie Apocalypse; The Rezort; Re-Kill; Maggie*

2016 *Isle Of The Dead; Train to Busan; Daylight's End; Viral; Pride And Prejudice And Zombies; The Autopsy of Jane Doe; Dead Rising: Endgame; Plan Z*

2017 *Anna and the Apocalypse; Day of the Dead: Bloodline; Trench 11; Zombies Have Fallen; The Cured*

2018 *Overlord; Deadsight; Patient Zero; Ever After; The Night Eats the World; Rampant; Cargo; Overlord; Office Uprising*

2019 *The Dead Don't Die; Blood Quantum; Zombieland: Double Tap; The Odd Family: Zombie on Sale; Rabid; Little Monsters; Zombie Tidal Wave; Shed of the Dead; Zombi Child; 3 from Hell*

2020 *Alone; Paradise Z; Train To Busan 2: Peninsula; Block Z*

2021 *Army of the Dead; Super Z; G Zombie*

SIR

Economic Modelling Specialist Intl. (EMSI) have rated the cities in the US and how they would be affected in a zombie outbreak. Cities classed as untenable (overcome) were New York, Tampa Florida, Southern California, and Chicago. Survivable areas include Boston, Salt Lake City, Columbus, Baltimore, Virginia Beach, and the infamous Denver airport USA (a place well worth researching if you really want to dive into this occult). In the UK I calculated the worst place for population overcome was London, Birmingham, Newcastle, Leeds, Manchester, Glasgow, Bradford.

The **SIR** model stands for:

Susceptible	speed of infection
Infectious	rates of infection
Recovered	rates of removal i.e., cure or death

People per square mile in Europe:

Monaco	51,519
Malta	1,642
Netherlands	1,316
Belgium	991
England	700
Germany	623
Czechia	359
Denmark	354
Italy	532

Poland	320
France	309
Wales	290
Portugal	288
Austria	281
Hungary	277
Spain	243
Romania	216
Greece	209
Scotland	168
Northern Ireland	147

The CDC mathematical breakdown include the following major implications for their discussed zombie apocalypse:

- Less dense
- More remote
- More protected by land/sea
- Less connected by transportation links, train lines, arterial links and airports

I used this algorithm to work out where the best possible places to be in terms of infection transmission rates in the world.

Discounted the top 25 countries on the international air index i.e., countries with extensive international flights. Then mathematically included locations that had natural expansive areas between populated towns, as low as 5 per mile with a 200-mile radius which took some time, but the results were as follows:

- Ulaanbaatar, Mongolia - population density 800/sq mi
- Perth, Australia - population density 864/sq mi
- Norilsk, (military city) Russia - population density 397/sq mi
- Baracoa, Cuba - population density 217/sq mi (fantastic unless zombies can swim from Florida)

Ouroboros

Cannibalism is the eating of the flesh of one's own species but technically if you eat yourself this is no longer the case. Fox News published an article on 19th November 2020 with the headline *"Grow-your-own human steaks meal kit is not 'technically' cannibalism, makers say"*. The company is called Ouroboros Steak (Uroborus) the ancient symbol of a serpent biting its tail *Oura*-tail and *Boros*-devourer so tail devourer. First seen in the 14th century BCE in Egypt. The god Ra and his union with Osiris in the underworld, this became associated with esoteric gnosticism, hermeticism and satanism.

In Roman mythology, the Ouroboros was associated with the god Saturn, because of time, connecting one year to the next, like the serpent swallowing its tail in iconography. Saturn swallowed his children, and with his scythe, symbolise the devouring of life or mortality. This is the origin of death being associated with aside the bringer of death.

Galatians 5:15 "*But if ye bite and devour one another, take heed that ye be not consumed one of another*". The company produced these kits where you can grow your own DNA in meat form and then consume it must have some people's tastebuds salivating.

'Cognitive dissonance' is a term that is increasingly being used for those who cannot see evidence when it is prima facie in front of an individual. I have read some studies this year that include another term called 'cognitive decline' for some vaccinated people who fall ill after the jab and become mentally isolated in their thought process. I believe we are in a spiritual war with the pharmaceutical industry and if you think this idea is a new one, you are wrong.

In 1917 an Austrian philosopher, educator and spiritualist called Rudolf Steiner wrote the following "*I have told you that the spirits of darkness are going to inspire their human hosts, in whom they will be dwelling, to find a vaccine that will drive all inclination toward spirituality out of people's souls when they are still very young, and this will happen in a roundabout way through the living body. Today, bodies are vaccinated against one thing and another; in future, children will be vaccinated with a substance which it will certainly be possible to produce, and this will make them immune, so that they do not develop foolish inclinations connected with spiritual life – 'foolish' here, or course, in the eyes of materialists. . . . ". . . a way will finally be found to vaccinate bodies so that these bodies will not allow the inclination toward spiritual ideas to develop, and all their lives people will believe only in the physical world they perceive with the senses. Out of impulses which the medical profession gained from presumption – oh, I beg your pardon, from the consumption [tuberculosis] they themselves suffered – people are now vaccinated against consumption, and in the same way they will be vaccinated against any inclination toward spirituality. This is merely to give you a particularly striking example of many things which will come in the near and more distant future in this field – the aim being to bring confusion into the impulses which want to stream down to earth after the victory of the [Michaelic] spirits of light in 1879".*

Ultimately eating human flesh and blood drinking by humans is now in the open and according to data on the increase. Commercially we are putting it in shoes, skateboards and given this with the fact that if water does in fact have memory and energy, our DNA is not being respected by humanity. Governments are talking about zombies; the DNA of humans is being altered (discussed elsewhere in this book) and those being vaccinated can no longer give their blood as a donor in some countries. Embryonic derivatives are being explored for our food, our drinks and the circumcised skin of baby boys is being used by women for beauty, as an elixir of life. If zombies are portrayed as loving blood and flesh, humans in their present existence are drowning in it, have a taste for it and that is worth considering.

"Pure-bloods" and "mudbloods" are terms presently entering the narrative which stems from the Harry Potter movies of wizards. The pure-blood means that they do not have any known muggles or muggle-born in their family tree (non-magical humans) and a *mud blood* is a foul name used for someone who is muggle-born (non-magic parents). I find these terms extremely uncomfortable for if it is found that the injectables have altered an individual's DNA and it has been done by deception and therefore no fault of their own. This could cause division among the populations of the Earth with even more divide and conquer while the purveyors of power view the chaos.

As you reach the end of this chapter, I suspect you feel this is miserable stuff. We were always told as children not to be afraid of the dead for it is those who are alive you should be careful of. In Bram Stoker's Dracula, vampires drink blood for Vit. D which is generated by sunlight as they cannot go into the sun themselves. It would seem at the time of authoring this book many health professionals are calling for people to get as much sunlight as they can, I hope you can see the correlation.

As we leave this uncomfortable chapter let me lighten the mood. I became aware of religion as a child while watching vampire movies, I remember one evening I asked my parents if vampires were afraid of crosses as a crucifix therefore it would not work on a bloodsucking Jew or Muslim. You would need the Star of David or a Rub-el-Hizb and with an atheist vampire you would be in real trouble... as you can imagine my parents did struggle with me.

I must defend Nosferatu or Dracula, for in the story telling world he is the greatest blood drinker of them all, remember the fact Dracula the bloodsucker did not actually kill anyone...well, not until he was dead.

Conclusion

We are all on a journey through life knowingly or unknowingly, unpredictable, unsuspected, and blind to what is going on around us. I have shown you a little of the mathematical matrix that sits behind everything we do and speak. Now turn this book upside down and read this paragraph again in fact read it twice.

You have just demonstrated how powerful the human mind is for you were reading right to left and upside down and the second time you did it faster. Let me demonstrate how occult magic can really hide within your mind's algorithms. Number rhythms exists in words you just do not see it, poets do, but how so? Single stanza is a tremendous example as a piece of work by Abdullah Shoaib. This numeric beat happens in the mind without you knowing. Settle yourself, your mind and read this poem aloud.

I'm very ugly
So don't try to convince me that
I am a very beautiful person
Because at the end of the day
I hate myself in every single way
And I'm not going to lie to myself by saying
There is beauty inside of me that matters
So rest assured I will remind myself
That I am a worthless, terrible person
And nothing you say will make me believe
I still deserve love
Because no matter what
I am not good enough to be loved
And I am in no position to believe that
Beauty does exist within me
Because whenever I look in the mirror I always think
Am I as ugly as people say?

What a negative poem I hear you cry but stop, wait, think, your take is completely wrong. How so? well look at the poem again but this time read it from the bottom line up, line by line...

The same words, the same lines but different viewpoints. One is you think you are ugly; the other is that you are beautiful. You just changed the pattern, perspective, and angle. Change the perspective and you change your frequency.

I believe letters have a spirit and spells; numbers have a soul. Mathematics is the language of the universe with music, words and thoughts which are the dialect that follows. We are in a temporary purgatory until we break out of the matrix through knowledge. Numbers do not weep, hope, bleed or feel. They know nothing of sacrifice, effort, trust, or love. The psychopath only ever considers ones and zeros, and bad politicians sacrifice the people for numbers, but great leaders sacrifice numbers for people. Humankind is creating more THINGS each day through invention, but we never stop to consider why the rush. From the second wheel to the fusion bomb humanity has a drive to create but does not question the direction, which results in discoveries far from the intention of the original quest. We fail to contemplate the possible result of our actions. We need to stop, smell the flowers, enjoy where we are, and not where we are going, look at what we are missing and what is being hidden or occulted from us.

When I wrote my first book, I had many criticisms thrown at me but to balance that so many more beautiful people told me *"thank you"*. As a builder, I create things. I arrive in a green space and when I leave there is a bridge, hospital, or home for people to keep warm and safe.

Be prolific, whether it be learning to play the violin, fly a plane, or swim in the sea, it is all about effort. Find your passion and invest your time, for every step you take you will get better. You must swim with the fishes to know the temperature of the water, only a dead fish will go with the flow, so find yourself. That inner voice which has an intuitive knowledge which rejects the social and cultural conditioning, the programmed mask which covers your true self. Stand, remove the mask and be seen.

Please hold this book in your hand knowing that I the author have written these pages in earnest and angst. When people of my generation are no longer walking the Earth, I believe humans will ask how did this happen? How did people fall for this madness? My answer to you is simple, there are very few of us who realised, and even fewer willing to fight. For those of you who have this book you can say you were a searcher of the truth, otherwise you would not be reading these words.

"Those who are able to see beyond the shadows and lies of their culture will never be understood, let alone believed, by the masses"- Plato

It is my calculation that 1% control the world. 4% are their puppets, to carry out their agenda. 90% of the population are asleep. This leaves 5% who understand this and are trying to wake the 90%. The 1% use the 4% to prevent the 5% from waking up the 90%. Therefore 4% with all the resources are fighting 5% that have none. Those who have refused the medical intervention will become the un-patented 1%. A statement worth pondering.

Numbers are the fingerprints from the blueprints of the universe. The question is who drew up the blueprints, and it follows *'who built what we see from those blueprints, and why'*?

So once again my fellow autodidactic, I hope you have enjoyed the beads I have left for you within these pages, and I truly thank you for your time, care, and courtesy. I am just a builder, but I am driven to write these pages and I intend to write more in continuance of the occult, on the thousands of others I know.

I hope we will do more of this again...for deep conversation. But for now, we must end our little journey...just you and me...with what time you could spare...here and there.

And remember a pizza holds the secrets to the universe.

With Love ... Namaste... gary... just a builder.

Gematria Keys

English Gematria or English Ordinal
A=1 B=2 C=3 D=4 E=5 F=6 G=7 H=8 I=9 J=10 K=11 L=12 M=13 N=14 O=15 P=16 Q=17 R=18 S=19 T=20 U=21 V=22 W=23 X=24 Y=25 Z=26

Reverse Ordinal
A=26 B=25 C=24 D=23 E=22 F=21 G=20 H=19 I=18 J=17 K=16 L=15 M=14 N=13 O=12 P=11 Q=10 R=9 S=8 T=7 U=6 V=5 W=4 X=3 Y=2 Z=1

Full Reduction
A=1 B=2 C=3 D=4 E=5 F=6 G=7 H=8 I=9 J=1 K=2 L=3 M=4 N=5 O=6 P=7 Q=8 R=9 S=1 T=2 U=3 V= 4 W=5 X=6 Y=7 Z=8

Elizabethan or Bacon Cypher
A=1 B=2 C=3 D=4 E=5 F=6 G=7 H=8 I=9 J=9 K=10 L=11 M=12 N=13 O=14 P=15 Q= 16 R=17 S=18 T=19 U=20 V=20 W=21 X=22 Y=23 Z=24

Bacon Kaye Cypher
A=27 B=28 C=29 D=30 E=31 F=32 G=33 H=34 I=35 J=35 K=10 L=11 M=12 N=13 O=14 P=15 Q=16 R=17 S=18 T=19 U=20 V=20 W=21 X=22 Y=23 Z=24

Sumerian Gematria
A=6 B=12 C= 18 D =24 E=30 F=36 G=42 H=48 I=54 J=60 K=66 L=72 M=78 N=84 O=90 P=96 Q=102 R=108 S=114 T=120 U=126 V=132 W=138 X=144 Y=150 Z=156

English Sumerian
A=156 B=150 C=144 D=138 E=132 F=126 G=120 H=114 I=108 J=102 K=96 L=90 M=84 N=78 O=72 P=66 Q=60 R=54 S=48 T=42 U=36 V=30 W=24 X=18 Y= 12 Z= 6

Satanic Gematria
A=36 B=37 C=38 D= 39 E=40 F=41 G=42 H=43 I=44 J=45 K=46 L=47 M=48 N=49 O=50 P=51 Q=52 R=53 S=54 T=55 U=56 V=57 W=58 X=59 Y=60 Z=61

Genesis Order
ב Beth = 1, א Aleph = 2, ג Gimel = 3, ש Shin = 3, ד Daleth = 4, ת Tav = 4, ה Heh = 5, ו Vav = 6, ז Zayin = 7, ח Cheth = 8, ט Teth = 9, י Yod = 10, כ Kaph = 11, ל Lamed = 12, מ Mem = 13, נ Nun = 14, ס Samekh = 15, ע Ayin = 16, פ Peh = 17, צ Tsade = 18, ק Qoph = 19, ר Resh = 20.

Table of ASCII values – American standard code for information interchange
A=65, B= 66, C= 67, D=68, E=69, F =70, G =71, H =72, I =73, J =74, K =75, L =76, M =77, N= 78, O=79, P=80, Q=81, 0=82, S=83, T=84, U=85, V=86, W=87, X=88, Y=89, Z=90

Binary Communications Chart

Character	Binary Code	Character	Binary Code	Character	Binary Code	Character	Binary Code	Character	Binary Code
A	01000001	Q	01010001	g	01100111	w	01110111	-	00101101
B	01000010	R	01010010	h	01101000	x	01111000	.	00101110
C	01000011	S	01010011	i	01101001	y	01111001	/	00101111
D	01000100	T	01010100	j	01101010	z	01111010	0	00110000
E	01000101	U	01010101	k	01101011	!	00100001	1	00110001
F	01000110	V	01010110	l	01101100	"	00100010	2	00110010
G	01000111	W	01010111	m	01101101	#	00100011	3	00110011
H	01001000	X	01011000	n	01101110	$	00100100	4	00110100
I	01001001	Y	01011001	o	01101111	%	00100101	5	00110101
J	01001010	Z	01011010	p	01110000	&	00100110	6	00110110
K	01001011	a	01100001	q	01110001	'	00100111	7	00110111
L	01001100	b	01100010	r	01110010	(	00101000	8	00111000
M	01001101	c	01100011	s	01110011	)	00101001	9	00111001
N	01001110	d	01100100	t	01110100	*	00101010	?	00111111
O	01001111	e	01100101	u	01110101	+	00101011	@	01000000
P	01010000	f	01100110	v	01110110	,	00101100	_	01011111

https://www.sciencefriday.com/educational-resources/write-your-name-in-binary-code/

Further Research

Project Bluebird, Project Blue Beam, Project Artichoke, Project Monarch, Project Rainbow, Project Woodpecker, Stargate Project, Project Blue Book, Project Coast, MKUltra, Project MKNAOMI, Operation CHAOS, Operation Gladio, Operation Mockingbird, Operation Paperclip, Operation Northwards, Operation Ranch Hand, Operation Popeye, Project Seal, Stargate Project, Operation Highjump, Operation Delirium, Operation Midnight Climax, Operation Cloverleaf, Operation Fishbowl, COINTELPRO, Project RAINBOW, Grill Flame Protocol, Project Sun Streak.

21st October 2010, the Federation of American Scientists made a statement that 5,135 inventions have been made secret by the US under the Invention Secrecy act of 1951. The suppression of these patents is a demonstration of how humans are trying to make better the world, but the cabal suppresses these ideas for its own gains.

One such subject is these Geoengineering Patents:

837 3962	011 7003	807 9545	765 5193
764 5326	801 2453	804 8309	655 3849
689 0497	005 6705	656 9393	796 5488
653 9812	653 9812	6412416	640 8704
638 2526	631 5213	628 1972	626 3744
611 0590	602 5402	6045089	603 4073
603 0506	591 2396	698 4239	5949001
592 2976	563 1414	580 0481	576 2298
563 9441	544 1200	562 8455	555 6029
548 6900	536 0162	543 4667	542 5413
538 3024	528 6979	535 7865	532 7222
529 6910	517 4498	524 5290	524 2820
514 8173	505 9909	515 6802	511 0502
510 4069	503 8664	505 6357	504 1834
504 1760	605 6203	500 5355	500 3186
499 9637	133 8343	494 8257	487 3928

It is no wonder humanity is not improving.

The author

Gary Fraughen holds a First-Class science degree in surveying, is Chartered and has won countless awards. He has worked in the construction industry all his life. Having taken expert witness roles for court proceedings an interest in language and etymology was sparked. Following the successful sale of his first book 'putting a stopper in the bottle of death and the occult' he continues his deep dive writing exploring the hidden nature of words and now numbers.

With Deepest Thanks
This book is dedicated to Aurelia, Michelle, Jay, Stuart and Mansoor without whose help this book would not have made it to print.